MULTIPLE CHOICE QUESTIONS TO PREPARE FOR THE AP CALCULUS BC EXAM

(second edition)

By Rita Korsunsky

www.mathboat.com

PREFACE

I am delighted to offer this book, titled "MULTIPLE CHOICE QUESTIONS TO PREPARE FOR THE AP CALCULUS BC EXAM".

This book consists of two parts. The first part contains six sample multiple-choice exams. Section I Part A consists of 30 multiple-choice questions to be answered without the use of a calculator. Section 1 part B consists of 15 multiple-choice questions where a graphing calculator could be used. The second part of this book contains a comprehensive list of definitions, formulas and theorems, and tips for the AP test.

The sole purpose of this book is to help you prepare for the AP Calculus BC exam. On the exam, you will have to respond to both multiple-choice and free-response questions. It is my belief that extensive practice with multiple-choice questions will help to prepare you well for both sections.

Do not forget to practice for the free response part of the test. I strongly suggest that you practice by answering the free response questions from past AP tests. These are available for free on the College Board's website.

Completing the multiple choice tests offered in this book will enable you to learn your strengths and weaknesses. You will find out what material you need to review. If you focus on learning, re-learning, and reviewing this material, you will greatly improve your performance on the test.

There is no substitute for practice. Taking multiple practice tests will teach you what your problem areas are, and will increase your confidence on the day of the exam.

I would like to thank all of my past, present, and future students for inspiring me to publish this Book. A special thanks to Lillian Li for hand drawing the exquisite book cover and Travis Chen for helping with publishing procedure. I also wish to thank my sons David and Boris for their technical assistance and support. A special debt of gratitude is due to my husband Alex for his continual support and help every step of the way.

Please send all questions and concerns to:

captain@mathboat.com

Sincerely,

Rita Korsunsky

If you wish to order the solutions with step-by-step explanations to each and every problem made in a form of PowerPoint presentation, please visit www.mathboat.com

www.mathboat.com

TABLE OF CONTENTS

Examination I……………………………………… 1

Examination II………………………………….... 24

Examination III…………………………………… 47

Examination IV…………………………………… 70

Examination V…………………………………… 93

Examination VI………………………………..... 116

Formulas and Theorems……………………….. 139

Tips for the AP test………………………………. 149

Answer key………………………………………. 151

Examination I

Section I Part A

Directions: Solve each of the following problems, using the space provided. Choose the best answer. Do not spend too much time on any one problem. Calculators may NOT be used on this part of the exam.

In this test: Unless otherwise specified, the domain of a function is assumed to be the set of all real numbers x for which $f(x)$ is a real number.

1. What is the absolute maximum of the function $f(x) = \frac{1}{2}x^4 - \frac{4}{3}x^3 + 2$ on the interval $[-1, 3]$?

$(A)\,3$ $(B)\,\dfrac{23}{6}$ $(C)\,\dfrac{13}{2}$ $(D)\,\dfrac{17}{2}$

$f'(x) = 2x^3 - 4x^2$

$0 = 2x^3 - 4x^2 \qquad \frac{1}{2} + \frac{4}{3} + 2$

$= 2x^2(x-2)$

$f(-1) = \frac{23}{6}$

$f(0) = 2$

$f(2) = -2/3 \qquad x = 0, 2$

$f(3) = 13/2$

$13/2 = C$

Answer $13/2 = C$

2. $\displaystyle \lim_{x \to \pi} \frac{\cos^3\left(\dfrac{x}{2}\right)}{\sin x} =$

$(A)\,\dfrac{1}{2}$ $(B)\,0$ $(C)\,1$ $(D)\,\dfrac{2}{3}$

Answer _____

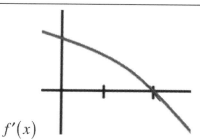

$f'(x)$

3. The graph of $f'(x)$ is shown above. The function $f(x)$ is twice differentiable. If $f(2)=4$, which of the following is true?

(A) $f''(2) < f(2) < f'(2)$

(B) $f''(2) < f'(2) < f(2)$

(C) $f'(2) < f''(2) < f(2)$

(D) $f(2) < f''(2) < f'(2)$

Answer _____ B _____

4. If the function f defined by $f(x) = \begin{cases} 1+e^{-2x}, & 0 \le x \le b \\ 1+e^{2x-12}, & b < x \le 6 \end{cases}$

is continuous for all values of x on the interval $[0,6]$, which of the following is the value of b?

(A) 1 (B) 2 (C) 3 (D) 4

Answer _____

5. Which of the three series below converge?

I. $\sum_{n=2}^{\infty} \frac{1}{n}$ II. $\sum_{n=2}^{\infty} \frac{\ln n}{n}$ III. $\sum_{n=2}^{\infty} \frac{(\ln n)^2}{n}$

(A) None (B) II only (C) III only (D) I and III

Answer_____

6. A rubber band in the shape of a circle is being evenly stretched, and its radius is increasing at a constant rate of 1.5 cm/sec. At what rate (in cm/sec) is the circumference of the rubber band changing when its radius is 6 centimeters?

(A) 2π

(B) 3π

(C) 6π

(D) 18π

Answer _____

Examination I

7. The graph of the function $f(x)$, consisting of two line segments, is shown in the figure below. Let g be the function given by $g(x) = x^2 + x - 1$, and let h be the function given by $h(x) = f(g(x))$. What is the value of $h'(1)$?

(A) 5 (B) $\dfrac{5}{3}$ (C) 3 (D) $\dfrac{3}{5}$

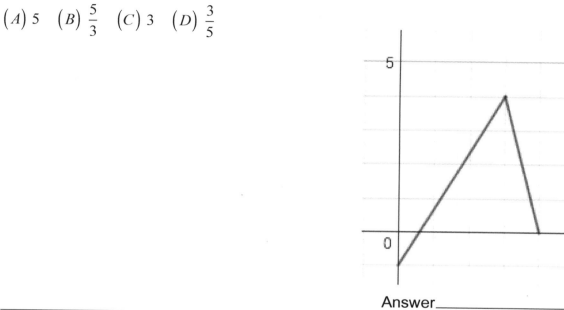

Answer_____

8. Given $f(x) = x^3 - 6x^2 + 12x - 8$ and $g(x)$ is its inverse, what is $g'(-8)$?

(A) $\dfrac{1}{8}$

(B) 12

(C) $\dfrac{1}{12}$

(D) 8

$x = y^3 - 6y^2 + 12y - 8$

$x + 8 = y^3 - 6y^2 + 12y$

$1 = 3y^2 y' - 12 y y' + 12 y' -$

Answer _____

4

9. The rate at which water is being pumped into a water tank is modeled by the function

$f(t) = -\frac{1}{4}t^3 + \frac{3}{2}t^2 + 3$. For the time interval $1 \le t \le 7$, at what time t is the water being pumped at the fastest rate?

$(A)\ t = 2$ $(B)\ t = 3$ $(C)\ t = 4$ $(D)\ t = 6$

Answer _____

10. Which of the following integrals is equal to the length of the curve $x = \dfrac{15y^3 - 2y - 1}{8}$ from

$A(5,\ 1.43)$ to $B(10,\ 1.78)$?

$(A) \displaystyle\int_{1.43}^{1.78} \sqrt{\frac{45y^2}{8} + \frac{3}{4}}\, dy$

$(B) \displaystyle\int_{1.43}^{1.78} \sqrt{1 + \left(\frac{45y^2}{8} - \frac{1}{4}\right)^2}\, dy$

$(C) \displaystyle\int_{5}^{10} \sqrt{1 + \left(15y^3 - 2y - 1\right)^2}\, dy$

$(D) \displaystyle\int_{1.43}^{1.78} \sqrt{1 + \left(\frac{45y^2}{8} - \frac{1}{2}\right)^2}\, dy$

Answer _____

11. A line perpendicular to the tangent line to the curve at the point of tangency is called a normal line to the curve at that point. Which of the following is the equation of the normal line to the curve $f(x) = \sqrt{x^3 + 3}$ at the point on the curve where $x = 1$?

$(A)\ 3y + 4x = 10$ $(B)\ 4y - 3x = 5$ $(C)\ 3y - 4x = 2$ $(D)\ 3y + 4x = 14$

Answer_____

12. $\displaystyle\int_{1}^{2}\left(1 + \frac{1}{x}\right)^{-2}\left(\frac{1}{x^2}\right)dx =$

$(A)\ \dfrac{1}{2}$ $(B)\ \dfrac{1}{6}$ $(C)\ \dfrac{1}{12}$ $(D)\ \dfrac{1}{3}$

Answer_____

13. The volume of the solid formed by revolving the region bounded by the graph of $2y = x^2 + 4$ and $y = x^2$ about $y = 4$ is given by which of the following integrals?

$(A) \pi \int_{-2}^{2} \left(2 - \frac{x^2}{2} \right)^2 dx$

$(B) \pi \int_{-2}^{2} \left(\left(x^2 \right)^2 - \left(\frac{x^2}{2} + 2 \right)^2 \right) dx$

$(C) \pi \int_{-2}^{2} \left(\left(4 - x^2 \right)^2 - \left(2 - \frac{x^2}{2} \right)^2 \right) dx$

$(D) \pi \int_{-2}^{2} \left(\left(4 - x^2 \right)^2 - \left(\frac{x^2}{2} \right)^2 \right) dx$

Answer_____

14. If $\lim\limits_{x \to 3} \dfrac{f(x)}{2x-6} = f'(3) = 0$, which of the following must be true?

(A) I only

(B) III only

(C) I and III only

(D) I, II, and III

I. $(3,0)$ is a critical point of $f(x)$

II. $f(x)$ has a local maximum at $x = 3$

III. $f(x)$ is continuous at $x = 3$

Answer _____

15. $\lim\limits_{h \to 0} \dfrac{\sin\left(\dfrac{\pi}{2}+h\right)-\sin\left(\dfrac{\pi}{2}\right)}{h}$ is

(A) $\dfrac{\sqrt{2}}{2}$

(B) 1

(C) 0

(D) nonexistent

Answer_____

16. $\displaystyle\int_{0}^{\pi} x\cos 3x\,dx =$

(A) $-\dfrac{2}{9}$

(B) 0

(C) $-\dfrac{2}{3}$

(D) $\dfrac{2}{9}$

Answer_____

17. Let the function $f(x)$ be continuos for all x. The figure below shows the graph of $f''(x)$. Which of the following statements about $f(x)$ or $f'(x)$ must be true?

I. f has a relative maximum in the open interval $c < x < d$.

II. f is concave up for $x < a$ and $b < x < c$.

III. f' has a relative minimum in the open interval $a < x < c$

(A) II only (B) III only (C) II and III only (D) I, II, and III

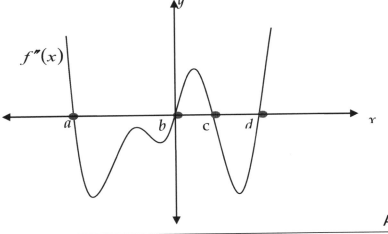

Answer_____

18. The convergent power series in x that has the sum equal to $\dfrac{x^3}{2-x^3}$ when $|x| < \sqrt[3]{2}$ is

$(A) \dfrac{1}{2} \displaystyle\sum_{n=0}^{\infty} \dfrac{x^{3n+3}}{2^n}$

$(B) \displaystyle\sum_{n=0}^{\infty} \dfrac{x^{3n+3}}{2^n}$

$(C) \dfrac{1}{2} \displaystyle\sum_{n=0}^{\infty} \dfrac{x^{3n}}{2^n}$

$(D) \dfrac{1}{2} \displaystyle\sum_{n=0}^{\infty} \dfrac{x^{3n+3}}{4^n}$

Answer _____

19. What is the slope of the tangent line to the curve $\left(x^3 + y\right)^2 - 2a^2xy = b^2$ at the point $P(0,1)$ when $a = \sqrt{2}$ and $b = 1$?

$(A) -2$

$(B) -\dfrac{1}{2}$

$(C)\ 2$

$(D)\dfrac{1}{2}$

Answer_____

20. A circle is inscribed in a square. The area of the circle is increasing at a constant rate of 15π in^2/sec. As the circle expands, the square expands to keep the circle inscribed. At what rate is the area of the square increasing in in^2/sec?

$(A)\ 30$ $(B)\ \dfrac{60}{\pi}$ $(C)\ \dfrac{120}{\pi}$ $(D)\ 60$

Answer _____

Examination I

21. A boy runs on a straight road for $0 \le t \le 8$ seconds. The graph below, which consists of two line segments, shows the velocity, in meters per second, of the boy. What is the total distance, in meters, run by the boy over the time interval $0 \le t \le 8$ seconds?

(A) 70

(B) 70.5

(C) 71

(D) 71.5

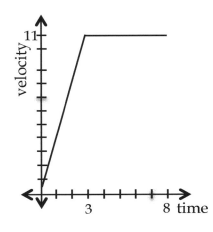

Answer_____

22. If n is a positive integer, then $\displaystyle \lim_{n \to +\infty} \frac{1}{n}\left[\frac{1}{1+\left(\dfrac{n+1}{n}\right)^2} + \frac{1}{1+\left(\dfrac{n+2}{n}\right)^2} + ... + \frac{1}{1+\left(\dfrac{n+n}{n}\right)^2} \right]$ could

be expressed as

$(A) \displaystyle\int_1^2 \frac{1}{x\left(x^2+1\right)}\, dx$

$(B) \displaystyle\int_1^2 \frac{1}{x^2+1}\, dx$

$(C) \displaystyle\int_0^2 \frac{1}{x\left(x^2+1\right)}\, dx$

$(D) \displaystyle\int_0^2 \frac{1}{x^2+1}\, dx$

Answer _____

23. The position of an object moving along the line at time t is described by the function $s(t) = -(t^2 - 2t + 4)(t^3 - 2t)$. What is the velocity of the object at time $t = 1$?

$(A) -3$

$(B) -1$

$(C) 1$

$(D) 3$

Answer_____

24. Shown on the right is the slope field for which differential equation?

$(A) \ \dfrac{dy}{dx} = 2x$

$(B) \ \dfrac{dy}{dx} = 2x - 4$

$(C) \ \dfrac{dy}{dx} = 4 - 2x$

$(D) \ \dfrac{dy}{dx} = x + y$

Answer _____

25. Consider the curve given by parametric equations: $x = 3t^4 - 2t^2$ and $y = t^4 - 5t$. What is $\dfrac{dy}{dx}$ at $t = 1$?

$(A) -\dfrac{2}{7}$ $(B) -\dfrac{1}{8}$ $(C) -\dfrac{7}{2}$ $(D) -8$

Answer_____

26. What are the interval(s) where the function $f(x) = \dfrac{8}{3}x^3 - 6x^2 - 36x + 7$ is increasing?

$(A)\ \left[-\dfrac{3}{2}, 3\right]$

$(B)\ (-\infty, 1]$ and $[3, \infty)$

$(C)\ \left(-\infty, -\dfrac{3}{2}\right]$ and $[3, \infty)$

$(D)\ \left(-\infty, -\dfrac{3}{2}\right]$ and $[1, \infty)$

Answer _____

27. Which of the following is the equation of the function $f(x)$ if its second derivative, $f''(x) = 2\sin x - 3\cos x$ and $f(0) = 7, f'(0) = 3$?

$(A)\ f(x) = 2\cos x + 3\sin x + 7x + 5$ $(B)\ f(x) = -2\sin x + 3\cos x + 5x + 4$

$(C)\ f(x) = -2\sin x + 3\cos x + 5x + 7$ $(D)\ f(x) = 3\sin x - 2\cos x - 3x + 9$

Answer_____

28. If $f'(x) = \lim\limits_{h \to 0} \dfrac{(x+h)^3 - x^3}{h}$, which of the following must be true:

I. $f'(x) \leq 0$ for all x

II. $f'(x) \geq 0$ for all x

III. $f(x)$ is concave down when $x \geq 0$

(A) I only

(B) II only

(C) III only

(D) II and III only

Answer_____

29. Using the table below, what is $\left(\dfrac{f}{g}\right)'(3)$?

(A) $-\dfrac{2}{3}$

(B) $-\dfrac{2}{27}$

(C) $\dfrac{2}{3}$

(D) $\dfrac{2}{27}$

$f(3)$	$g(3)$	$f'(3)$	$g'(3)$
7	-9	-4	6

Answer_____

30. The graph of f'', the second derivative of f, is the line shown in the figure below. If $f'(0) = 2.5$, then $f'(2) =$

(A) 9.5

(B) 10.5

(C) 11.0

(D) 11.5

$y = f''(x)$

Answer_____

Examination I

Section I Part B

Directions: Solve each of the following problems, using the space provided. Choose the best answer. Do not spend too much time on any one problem. A graphing calculator is required for some questions on this part of the exam.

In this test:

1. The exact numerical value of the correct answer does not always appear among the choices given. Then select from among the choices the number that best approximates the exact numerical value.

2. Unless otherwise specified, the domain of a function is assumed to be the set of all real numbers x for which $f(x)$ is a real number.

31. Which of the following is the radius of convergence of the series $\displaystyle\sum_{n=0}^{\infty} \frac{(-1)^n n! x^n}{n^n}$?

$(A) \dfrac{1}{e}$ $(B) 1$ $(C) e$ $(D) \infty$

_____ Answer_____

32. The composition of two functions, $h(x) = f(g(x))$ is a differentiable function. Using the tables below what is the approximate value of the derivative of $h(x)$ at $x = 1.35$?

$(A) 15.88$ $(B) 10.72$ $(C) 3.97$ $(D) 13.29$

x	2.5	2.7	2.9
$f'(x)$	3.85	3.97	4.03

x	1.30	1.35	1.40
$g(x)$	2.5	2.7	2.9

_____ Answer _____

33. If $f(2) = 8$ and $f'(x) = \dfrac{\cos(1-x^2)}{x^2 + \sqrt{x}}$, then $f(7) \approx ?$

(A) 8.170

(B) 0.170

(C) 7.982

(D) -8.018

Answer_____

34. What is the length of the curve given by the pair of parametric equations $x = \cos 3t$ and $y = \sin^3 t$ from $t = \pi$ to $t = 3\pi$?

(A) 12.513   (B) 13.262   (C) 14.563   (D) 15.827

Answer_____

35. The function $f(x) = \sin x$ is defined on the interval $[0, b]$. The line tangent to the graph of $f(x)$ at $x = 1.531$ is parallel to the segment connecting the endpoints of $[0, b]$. What is the value of b?

(A) 2.704

(B) 4.787

(C) 3.567

(D) 3.021

Answer_____

36. Water is being pumped continuously from a water pool at a rate proportional to the amount of water left in the pool; that is, $\dfrac{dy}{dt} = ky$, where y is the amount of water left in the pool at any time t. Initially there were 500,000 gallons of water in the pool, and 10 days later there were 100,000 gallons left. What is the equation for y, the amount of water remaining in the pool at any time t?

(A) $y(t) = 500{,}000e^{\frac{1}{5}t}$

(B) $y(t) = 500{,}000\left(\dfrac{1}{10}\right)^{\frac{t}{10}}$

(C) $y(t) = 500{,}000 \cdot \left(\dfrac{1}{5}\right)^{\frac{t}{10}}$

(D) $y(t) = 500{,}000e^{10t}$

Answer _____

37. Two bikers start at the same place and same time. The first biker rides east at a constant velocity of 5 miles per hour and the second biker rides north at a constant velocity of 12 miles per hour. Approximately how fast is the distance between them changing after 3 hours?

(A) 7mph

(B) 12mph

(C) 13mph

(D) 17mph

Answer_____

38. Let f be the function whose derivative is $f'(x) = \dfrac{1+e^x}{x^2}$ and whose graph passes through the point (3,6). What is the approximate value of $f(3.1)$ if $\displaystyle\int_{3}^{3.1} f'(x)\,dx = 0.2377$?

(A) 6.238 (B) 2.414 (C) 6.1 (D) -5.762

Answer _____

39. What is the average value of the function $G(x) = 30 - \dfrac{x^8}{2,500,000}$ on the interval where $G(x) \geq 0$?

(A) 28.473

(B) 27.134

(C) 26.667

(D) 25.431

Answer_____

40. The region bounded by the graphs of the equations $y = 5$ and $5x^4 - y = 3x^2 - 3$ is revolved around the x – axis. What is the approximate volume of the resulting solid?

(A) 31.416

(B) 68.793

(C) 81.274

(D) 98.696

Answer _____

41. If $f(x) = \int_{2}^{x^2} \sin t\, dt$ is continuous for all real numbers

and $g(x)$ is the derivative of $f(x)$, find $g'(5)$.

(A) 9.912 (B) 98.856 (C) 110.828 (D) 49.428

Answer_____

42. At which x-coordinate does the function $f(x) = 2\sin x + 4x$ have a point of inflection on the interval $(0, 2\pi)$?

(A) $\dfrac{\pi}{2}$

(B) $\dfrac{3\pi}{4}$

(C) $\dfrac{3\pi}{2}$

(D) π

Answer _____

43. Let R be the region that lies inside the polar curve $r = 2 - \cos\theta$ and also inside the polar curve $r = 1 + \cos\theta$. What is the area of R?

(A)1.611 (B)2.658 (C)2.457 (D)3.222

Answer_____

44. The base of a solid is the region in the fourth quadrant enclosed by the graph of $y = x^2 - 9$ and the coordinate axes. If every cross section perpendicular to the x-axis is a square, then the volume of the solid is

(A)64.8 (B)129.6 (C)171.0 (D)194.4

Answer_____

45. If $\int_1^7 f(x)\,dx = 6$ and $\int_7^1 g(x)\,dx = 12$ and $\int_1^7 h(x)\,dx = \int_1^5 \big(2f(x) - g(x)\big)\,dx + \int_7^5 \big(g(x) - 2f(x)\big)\,dx$,

what is the average value of $h(x)$ on the interval $1 \le x \le 7$?

$(A)\ 1$ $(B)\ 2$ $(C)\ 4$ $(D)\ 12$

Answer_____

Examination II

Section I Part A

Directions: Solve each of the following problems, using the space provided. Choose the best answer. Do not spend too much time on any one problem. Calculators may NOT be used on this part of the exam.

In this test: Unless otherwise specified, the domain of a function is assumed to be the set of all real numbers x for which $f(x)$ is a real number.

1. Let f be defined as follows, where $a \neq 0$.

$$f(x) = \begin{cases} \dfrac{x^3 - a^3}{x^2 - a^2}, & \text{for } x \neq a \\ 1, & \text{for } x = a \end{cases}$$

Which of the following must be true about f ?

       I. $\lim\limits_{x \to a} f(x)$ exists

       II. $f(a)$ exists

       III. $f(x)$ is continuous at $x = a$

(A) I only (B) II only (C) I and II only (D) I, II, and III

Answer_____

2. The slope of a curve at point (x, y) is defined as $\lim\limits_{h \to 0} \dfrac{(x+h)^3 + (x+h)^2 - x^3 - x^2}{h}$. Which of the following is the equation of the tangent to this curve at $x = 1$?

$(A)\, y = 5x - 2$

$(B)\, y = 3x - 9$

$(C)\, y = 5x - 3$

$(D)\, y = 3x - 6$

Answer _____

3. $\int x \cdot (2x^2 + 1)^4 dx =$

(A) $\frac{1}{4}(2x^2 + 1)^5 + C$

(B) $\frac{1}{4}(2x^2 + 1)^4 + C$

(C) $\frac{1}{20}(2x^2 + 1)^5 + C$

(D) $\frac{1}{20}(2x^2 + 1)^6 + C$

Answer_____

4. If $f(x) = g(h(x))$ and if $h(2) = 5$, $h'(2) = -5$, $g'(5) = 3$ which of the following is $f'(2)$?

(A) 3

(B) -15

(C) 15

(D) -3

Answer _____

5. Which of the following is y', the first derivative of the function $y = f(x)$ if $x^2 y + \sec y = 8$?

$(A)\ -2xy \cdot \left(x^2 \sec y \tan y\right)$ $(B)\ \dfrac{x^2 y}{\sec y \tan y}$ $(C)\ \dfrac{-2xy}{x^2 - \sec y \tan x}$ $(D)\ \dfrac{-2xy}{x^2 + \sec y \tan y}$

Answer_____

6. If $f(x) = 6x^2 + 4x$, what is the number c in the interval $[0,4]$ such that the line tangent to the graph of $f(x)$ at the point $x = c$ is parallel to the line drawn through the endpoints of the interval?

$(A)\ \dfrac{5}{3}$

$(B)\ 4$

$(C)\ \dfrac{3}{5}$

$(D)\ 2$

Answer _____

7. Let f be a function that is differentiable for all x. The value of $f'(x)$ is given for several values of x in the table below.

If $f'(x)$ is always decreasing, which of the following must be true?

I. $f(x)$ is an odd function.

II. $f(x)$ has a point of inflection at $\left(0, f(0)\right)$.

III. $f(x)$ has a relative maximum at $x = 0$.

(A) II only (B) III only (C) II and III only (D) I, II and III

x	-12	-6	0	6	12
$f'(x)$	5	2	0	-2	-5

Answer_____

8. $\int x^2 \sec^2\left(x^3\right) dx =$

$(A) \dfrac{1}{3}\tan^2\left(x^3\right) + C$

$(B) \, 3\tan\left(x^3\right) + C$

$(C) \, 3\sec\left(x^3\right) + C$

$(D) \dfrac{1}{3}\tan\left(x^3\right) + C$

Answer _____

9. For which of the following values of k do both $\sum_{n=1}^{\infty}\left(\dfrac{3}{k}\right)^n$ and $\sum_{n=1}^{\infty}\dfrac{(3-k)^n}{\sqrt{n+3}}$ converge?

(A) 2

(B) 3

(C) 4

(D) 5

Answer_____

10. Let f be the function given by $f(x)=e^{2x}$ for $-\infty<x\le 0$ and let R be the region between the graph of f and the x-axis. Which of the following is equal to the area of the region R?

(A) $\dfrac{1}{4}$

(B) $\dfrac{1}{2}$

(C) 1

(D) The area is infinite

Answer _____

11. Let $g(x)$ be the inverse of $f(x)$. The selected values of $f(x)$ and $f'(x)$ are given in the table below. Which of the following is the value of $g'(3)$?

(A) $\dfrac{1}{8}$

(B) $\dfrac{1}{7}$

(C) 3

(D) $\dfrac{1}{5}$

x	0	1	3	5
$f(x)$	-2	3	1	2
$f'(x)$	7	5	8	4

Answer_____

12. The slope of function $y = f(x)$ is given by $6x^4 - 12x^2 + 2$. What are the x-coordinates of all points of inflections of $y = f(x)$?

(A) $-\dfrac{\sqrt{3}}{3}, \dfrac{\sqrt{3}}{3}$

(B) $-1, \dfrac{\sqrt{3}}{3}, 1$

(C) $0, 1$

(D) $-1, 0, 1$

Answer _____

13. The Taylor series about $x = 2$ for a certain function f converges to $f(x)$ for all x in the interval of convergence. Then nth derivative of f at $x = 2$ is given by $f^{(n)}(2) = \dfrac{2^n n!}{3^{n+1}(n+1)}$ and $f(2) = 3$.

Which of the following is the radius of convergence for the Taylor series for f about $x = 2$?

(A) $\dfrac{2}{3}$ (B) 0   (C) $\dfrac{3}{2}$ (D) 3

_____ Answer _____

14. The volume of the solid formed by revolving the region bounded by the graphs of $y = x^2 + 2$, $y = x + 1$, $x = 0$, and $x = 1$ about the horizontal line $y = 4$ is given by which of the following ?

(A) $\pi \displaystyle\int_0^1 \left((3 - x)^2 - (2 - x^2)^2 \right) dx$

(B) $\pi \displaystyle\int_0^1 \left((x + 1)^2 - (x^2 + 2)^2 \right) dx$

(C) $\pi \displaystyle\int_0^1 \left((2 - x^2)^2 - (3 - x)^2 \right) dx$

(D) $\pi \displaystyle\int_0^1 \left((3 - x)^2 - (6 - x^2)^2 \right) dx$

_____ Answer _____

15. Shown below is the slope field of the differential equation. What could be the solution to this differential equation?

$(A) y = x^3$

$(B) y = -5x^2$

$(C) y = x$

$(D) y = x^2$

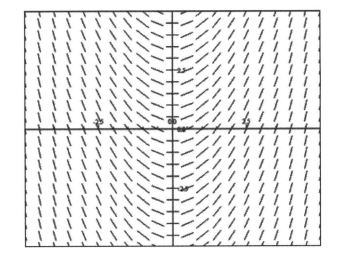

Answer_____

16. Which of the following is the derivative of $\sin^4\left(\cot^3 7x\right)$?

$(A)\ 84\sin^3\left(\cot^3 7x\right)\cot^2 7x$

$(B)\ -84\sin^3\left(\cot^3 7x\right)\cos\left(\cot^3 7x\right)\csc^2 7x$

$(C)\ -84\sin^3\left(\cot^3 7x\right)\cos\left(\cot^3 7x\right)\cot^2 7x\csc^2 7x$

$(D)\ 12\sin^3\left(\cot^3 7x\right)\cos\left(\cot^3 7x\right)\cot^2 7x\csc^2 7x$

Answer_____

17. The function $f(x) = x^4 + 3x^3 + 2x + 4$ must have a zero between which of the following values of x ?

$(A) -2$ and 1

$(B) 1$ and 2

$(C) 2$ and 3

$(D) 3$ and 4

Answer_____

18. $\lim\limits_{h \to 0} \dfrac{\cos\left(\dfrac{\pi}{3} + h\right) - \dfrac{1}{2}}{h} =$

$(A)\ \sqrt{3}$ $(B) -\dfrac{\sqrt{3}}{2}$ $(C)\ -\dfrac{1}{2}$ $(D)\ \dfrac{\sqrt{3}}{2}$

Answer _____

19. What are y coordinates of the points where $y^3 + x^2 + 30y^2 = 4x + 3$ has vertical tangent lines?

(A) $y = 0$ (B) $y = -20$ (C) $y = 0$ and $y = -20$ (D) $y = -20$ and $y = 2$

Answer_____

20. The region is bounded by the curve $f(x) = \dfrac{1}{(x+1)(4-x)}$ and $x-$axis from $x = 0$ to $x = 3$.

Which of the following is equal to the area of this region ?

$(A)\dfrac{2\ln2}{5}$ $(B)\dfrac{6\ln2}{5}$ $(C)\dfrac{4\ln2}{5}$ $(D)\dfrac{8\ln2}{5}$

Answer _____

Examination II

21. Let $f(x)$ be a differentiable function defined on the interval $-6 \le x \le 6$. The values of $f(x)$ and its derivative $f'(x)$ at several points of the domain are shown on the table below. The equation of the line tangent to the graph of $f(x)$ and parallel to the segment between the endpoints of $f(x)$ is

(A) $y=-x+3$ (B) $y=-3x+7$ (C) $y=-3x+4$ (D) $y=-4x+9$

x	-6	-4	-2	0	2	4	6
$f(x)$	41	25	13	3	2	4	5
$f'(x)$	-5	-4	-3	-1	0	1	0

Answer_____

22. Which of the following graphs is the graph of a differentiable function $f(x)$ that satisfies the given conditions:

$f(3)=2, f'(0)=f'(3)=f'(5)=0,$
$f'(x)<0$ if $x<0$ or $3<x<5$?

(A)

(B)

(C)

(D)

Answer_____

23. Function $f(x)$ is a linear function such that $f(x+2) - f(x) = 8$. Given that $g(x)$ is the inverse of $f(x)$, which of the following is $g'(x)$?

$(A)\ \dfrac{1}{8}$

$(B)\ \dfrac{1}{4}$

$(C)\ \dfrac{1}{2}$

$(D)\ 4$

Answer_____

24. $\displaystyle\lim_{x \to 3} \dfrac{\displaystyle\int_{3}^{x} \cos\left(\dfrac{\pi t}{3}\right) dt}{x^3 - 27} =$

$(A)\ 0$

$(B)\ -\dfrac{1}{27}$

$(C)\ -3$

$(D)\ \dfrac{1}{27}$

Answer _____

25. The graph of $f'(x)$ shown has horizontal tangents at $(4,0)$ and $(12,-6)$. What are the x-coordinates for the point(s) of inflection of function $f(x)$?

(A) 4 only (B) 4 and 15 only (C) -3 and 20 only (D) 4 and 12 only

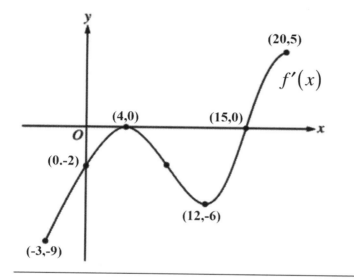

Answer_____

26. Let f be a function having derivatives of all orders for all real numbers. The third-degree Taylor polynomial for f about $x = 2$ is given by $T(x) = 5 - 9(x-2)^2 - 3(x-2)^3$. $\left|f^{(4)}(x)\right| \le 9$ for all x in the interval $[0,2]$. Which of the following must be true?

I. The approximation for $f(0)$ using $T(x)$ is negative.

II. f has a critical point at $x = 0$.

III. $f(0)$ is negative.

(A) I only

(B) III only

(C) I and II only

(D) I and III only

Answer _____

27. If n is a positive integer, then $\displaystyle\lim_{n\to\infty}\frac{1}{n}\left[\cos\frac{\pi}{n}+\cos\frac{2\pi}{n}+...+\cos\frac{n\pi}{n}\right]$ can be expressed as

$(A)\displaystyle\int_0^1\cos\frac{\pi}{n}dx$ $(B)\pi\displaystyle\int_0^1\cos x\,dx$ $(C)\displaystyle\int_0^1\cos(\pi x)dx$ $(D)\displaystyle\int_0^\pi\cos\pi x\,dx\,?$

Answer_____

28. Given the function $f(x)=3x^3+ax^2+bx+k$ where a, b, and k are constants. This function has a local minimum at $x=-2$ and a point of inflection at $x=-3$. What is the value of b?

$(A)\ 81$

$(B)\ 72$

$(C)\ 54$

$(D)\ 27$

Answer _____

29. The acceleration of a particle moving along a horizontal line at time t is $2\tan t \sec^2 t$. If the initial velocity is 1 and the initial position is 0, what is the position of the particle at time $t = \dfrac{\pi}{4}$?

$(A) -1$ $(B) 1$ $(C) 0$ $(D) 2$

Answer_____

30. A solid has a circular base of radius 4. If every plane cross section perpendicular to the x-axis is an equilateral triangle, then the volume is

$(A) \dfrac{256}{3}$

$(B) \dfrac{64}{3}\sqrt{3}$

$(C) \dfrac{128}{3}\sqrt{3}$

$(D) \dfrac{256}{3}\sqrt{3}$

Answer_____

Examination II

Section I Part B

Directions: Solve each of the following problems, using the space provided. Choose the best answer. Do not spend too much time on any one problem. A graphing calculator is required for some questions on this part of the exam.

In this test:

1. The exact numerical value of the correct answer does not always appear among the choices given. Then select from among the choices the number that best approximates the exact numerical value.

2. Unless otherwise specified, the domain of a function is assumed to be the set of all real numbers x for which $f(x)$ is a real number.

31. Consider the curves $r = 3\sin\theta$ and $r = 1 + \sin\theta$. Which of the following is the area of the region inside the curve $r = 3\sin\theta$ and outside the curve $r = 1 + \sin\theta$?

$(A)\ \dfrac{\pi}{2}$ $(B)\ \dfrac{2\pi}{3}$ $(C)\ \pi$ $(D)\ 2\pi$

Answer_____

32. If f is an antiderivative of $\dfrac{\sin^2 x}{x^2 + 2}$ such that $f(2) = \dfrac{1}{2}$, then $f(0) = ?$

$(A)\ -0.325$

$(B)\ 0.175$

$(C)\ 0.825$

$(D)\ 1.175$

Answer_____

33. The figure below shows the graph of f', the derivative of the function f, on the closed interval $-1 \le x \le 5$. The graph of f' has horizontal tangent lines at $x = 1$ and $x = 3$. The function f is twice differentiable with $f(3) = 8$. Let g be the function defined by $g(x) = x^2 f(x)$. Which of the following is the equation for the line tangent to the graph of g at $x = 3$?

$(A)\ y - 24 = 72(x + 3)$

$(B)\ y - 9 = 21(x - 3)$

$(C)\ y - 72 = 21(x - 3)$

$(D)\ y - 36 = 14(x - 3)$

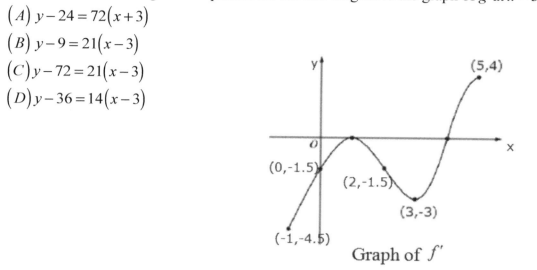

Graph of f'

Answer_____

34. The position $s(t)$ of a point P moving along a line is given by the function $s(t) = 5t^3 - 2t^2 + 8$ with t in seconds and $s(t)$ in centimeters. What is the average velocity in cm/sec of point P on the time interval $[1, 1.4]$?

$(A)\ 32.00$

$(B)\ 34.93$

$(C)\ 17.00$

$(D)\ 16.00$

Answer _____

Examination II

35. Let f be the function defined by $f(x) = \begin{cases} \sqrt{x+2} & \text{for } 0 \le x \le 2 \\ 4-x & \text{for } 2 < x \le 5.5 \end{cases}$

What is the average value of $f(x)$ on the closed interval $0 \le x \le 5.5$?

(A) 0.786 (B) 0.891 (C) 1.043 (D) 0.915

Answer_____

36. The rodent population in a village increases according to the equation $P(t) = 400 - 250e^{-0.123t}$
for $t \ge 0$, t measured in months. This population will approach a limiting value as time goes on.
During which month will the population reach a half of its limiting value?

(A) First (B) Second (C) Third (D) Fourth

Answer _____

37. The velocity vector $v(t)$ of a particle moving in the xy-plane is given by $(-4\cos t, -5\sin t)$ for $t \geq 0$. At $t = 0$, the particle is at the point $(1,3)$. Which of the following is the position vector at $t = 4$?

$(A)(-2.027, 4.732)$

$(B)(4.027, -5.268)$

$(C)(-3.027, 3.268)$

$(D)(4.027, -1.268)$

Answer_____

38. A graph of a function consists of a line segment from the point $(0,10)$ to point $(8,8)$, another line segment from $(8,8)$ to $(12,8)$, and a third line segment from $(12,8)$ to $(20,0)$. What is the average value of this function on the interval $[0,20]$?

(A) 136

(B) 7

(C) $\dfrac{34}{5}$

(D) $\dfrac{127}{20}$

Answer _____

39. A reactor coolant tank is being filled with coolant at the rate of $400\sqrt[3]{t}$ gallons per hour with $t > 0$ measured in hours. If the tank originally contained 200 gallons of coolant, how many gallons are in the tank after 8 hours?

(A) 800

(B) 4800

(C) 5000

(D) 6400

Answer _____

40. Let P be the partition of an interval $[2,6]$ determined by $\{2,4,5,6\}$. Using the Midpoint Riemann Sum, which of the following is the best approximation for the area under the curve of $f(x) = x^2 + 4$ on the interval $[2,6]$?

(A)65 (B)75.25 (C)55 (D)84.5

Answer _____

41. A particle, initially at rest, moves along the x – axis so that its acceleration at any time is given by $a(t) = 8t^2 - 6$. Which of the following is the total distance traveled by the particle within the time interval $2 \leq t \leq 4$?

(A) $\dfrac{412}{3}$

(B) 148

(C) $\dfrac{340}{3}$

(D) 124

Answer_____

42. On the diagram below, the cross section of a water tank that is being drained is an equilateral triangle. The area of the cross section of water, $A(t)$, is decreasing at a rate of $10 \text{cm}^2/\text{min}$. Which of the following is the rate (in cm/min) at which length of its side s is changing when area of the triangle is 250 cm^2?

(A) -0.240 (B) -0.481 (C) -0.832 (D) -0.951

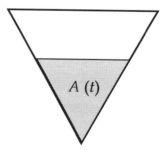

Answer_____

43. Let f be the function given by $f(t) = \dfrac{6}{1+t^3}$. Let G be the function given by $G(x) = \int_0^x f(t)\,dt$.

Which of the following are the first 4 nonzero terms for the Taylor series expansion of $G(x)$

about $x = 0$?

(A) $-3x^2 + 6x^5 - 9x^8 + 27x^{11}$

(B) $-18x^2 + 36x^5 - 54x^8 + 72x^{11}$

(C) $6x - \dfrac{6x^4}{4} + \dfrac{6x^7}{7} - \dfrac{6x^{10}}{10}$

(D) $6 - 6x^3 + 6x^6 - 6x^9$

Answer _____

44. A pizza, heated to a temperature of 475 degrees Fahrenheit $(°F)$, is taken out of an oven and

placed in a 105°F room at $t = 0$ minutes. The temperature of the pizza is changing at a rate of

$-256e^{-0.7t}$ degrees Fahrenheit per minute. To the nearest degree, what is the temperature of the

pizza at $t = 9$ minutes?

(A) 100 (B) 110 (C) 115 (D) 120

Answer _____

45. The region bounded by the graphs of the equations $\left(\dfrac{x}{2}\right)^2 - y = -2$ and $y = \dfrac{x}{5} + 4$ and by the vertical lines $x = 1$ and $x = 3$ is revolved around the x-axis. What is the volume of the resulting solid?

(A) 38.110

(B) 45.312

(C) 59.863

(D) 64.822

Answer_____

Examination III

Section I Part A

Directions : Solve each of the following problems, using the space provided. Choose the best answer. Do not spend too much time on any one problem. Calculators may NOT be used on this part of the exam.

In this test : Unless otherwise specified, the domain of a function is assumed to be the set of all real numbers x for which $f(x)$ is a real number.

1. Let f be defined as follows,

$$f(x) = \begin{cases} x^2 + 1, & \text{for } -2 \le x < 2 \\ 1, & \text{for } x = 2 \\ 5, & \text{for } 2 < x < \infty \end{cases}$$

What is the limit of $f(x)$ as x approaches 2?

(B) 1 (C) 2 (C) 5 (D) nonexistant

Answer_____

2. What is the number represented by the difference between the average rate of change of function $y = f(x)$ with respect to x on the given interval $[1, 1.5]$ and the instantaneous rate of change of y with respect to x at the right endpoint of this interval if $y = 4x^2 - 1$?

(A) 1

(B) 2

(C) 6

(D) 8

Answer_____

3. Which of the following is equal to the area of the region bounded by the lines $x = -3, x = 1$, $y = 0$ and the curve $y = \dfrac{x+22}{x^2+2x-8}$ graphed below ?

$(A)\ln 5$ $(B)6\ln 5$ $(C)7\ln 5$ $(D)8\ln 5$

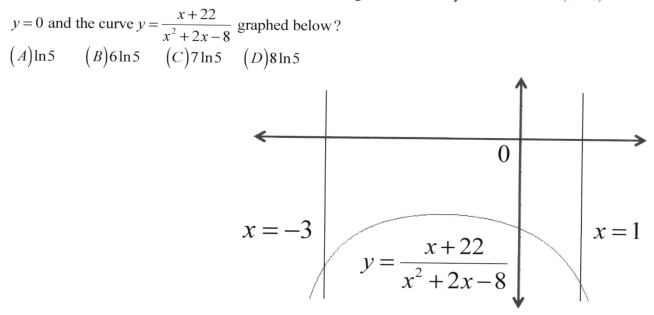

Answer_____

4. Given $xy^2 = 4$, what is the value of the second derivative of the function $y = f(x)$ at the point $(1,2)$?

$(A)\ \dfrac{3}{2}$

$(B)\ \dfrac{3}{4}$

$(C)\ 1$

$(D)\ \dfrac{1}{2}$

Answer_____

5. What is y if $\dfrac{dy}{dx} = \dfrac{3x+2}{5y}$ and $y = 1$ when $x = 2$?

$(A)\, y = \sqrt{\dfrac{3x^2 + 4x - 15}{5}}$ $\qquad (B)\, y = \pm\sqrt{\dfrac{3x^2 + 4x - 15}{5}}$

$(C)\, y = \pm\sqrt{\dfrac{3x^2 + 4x + 13}{2}}$ $\qquad (D)\, y = \sqrt{\dfrac{3x^2 + 4x}{5}}$

Answer_____

6. Which of the following definite integrals gives the length of the graph $y = e^{2x}$ between $x = 0$ and $x = 3$?

$(A)\, \displaystyle\int_0^3 \sqrt{1 + 4e^{4x}}\, dx$

$(B)\, \displaystyle\int_0^3 \sqrt{1 + e^{4x}}\, dx$

$(C)\, \displaystyle\int_0^3 \sqrt{1 + e^{2x}}\, dx$

$(D)\, \displaystyle\int_0^3 \sqrt{1 + 2e^{2x}}\, dx$

Answer_____

7. The figure below shows the graph of the derivative of a function f. How many points of inflection does f have in the interval shown on the diagram?

(A) Zero

(B) One

(C) Two

(D) More than two

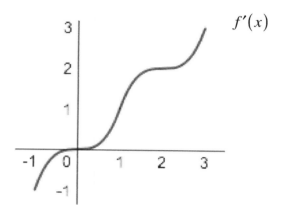

Answer_____

8. Which of the following is the maximum area of the rectangle that can be inscribed in a semicircle of radius $4\sqrt{2}$, if two vertices lie on the diameter?

(A) 24

(B) 32

(C) 36

(D) 64

Answer _____

9. Which of the following improper integrals diverge?

I. $\int\limits_{1}^{\infty} \dfrac{1}{x^{\frac{1}{3}}} dx$ II. $\int\limits_{0}^{\infty} \dfrac{1}{(x+2)^{2}} dx$ III. $\int\limits_{0}^{\infty} \dfrac{1}{e^{4x}} dx$

(A) I only

(B) II and III only

(C) I and II only

(D) I, II, and III

Answer_____

10. $\int\limits_{4}^{6} |x-5| dx =$

$(A)\, 4$

$(B)\, 0$

$(C)\, 1$

$(D)\, 2$

Answer _____

11. $\int e^{-x}\cos x\,dx =$

$(A)\dfrac{1}{2}e^{-x}\left(\sin x-\cos x\right)+C$

$(B)\dfrac{1}{2}e^{-x}\left(\sin x+\cos x\right)+C$

$(C)-\dfrac{1}{2}e^{-x}\left(\sin x+\cos x\right)+C$

$(D)\dfrac{1}{2}e^{-x}\left(\cos x-\sin x\right)+C$

Answer_____

12. The graph of $f(x)$ is shown below. List the following values of $f'(1), f'(-2), f''(0)$ from smallest to largest.

$(A)\, f'(-2),\, f''(0),\, f'(1)$
$(B)\, f''(0),\, f'(1),\, f'(-2)$
$(C)\, f''(0),\, f'(-2),\, f'(1)$
$(D)\, f'(1),\, f''(0),\, f'(-2)$

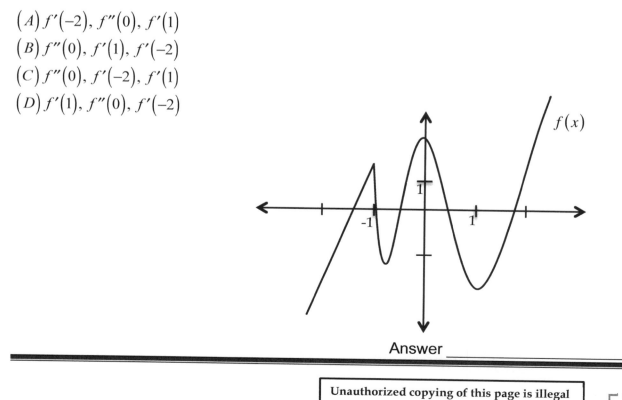

Answer _____

13. $\int \dfrac{dx}{1+\sin x} =$

(A) $-\dfrac{1}{\left(1+\sin x\right)^2}+C$

(B) $\tan x + \sec x + C$

(C) $\ln\left|1+\sin x\right|+C$

(D) $\tan x - \sec x + C$

Answer_____

14. The function $h(x)=\dfrac{f(x)}{g(x)}$ where $f(x)$ and $g(x)$ are piecewise linear functions whose graphs

are shown below. What is $h'(-2)$?

$(A) -\dfrac{3}{2}$ $(B) -1$ $(C) -\dfrac{1}{2}$ $(D) 1$

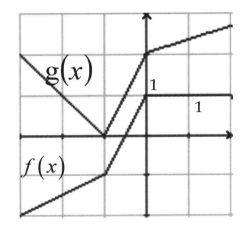

Answer _____

15. The graph of $f(x)$ is the semicircle with radius 5 as shown below. What is the mean value of $f(x)$ on the interval $[0,10]$?

$(A)\dfrac{5}{4}\pi$

$(B)\dfrac{5}{2}\pi$

$(C)5\pi$

$(D)\dfrac{25\pi}{2}$

Answer_____

16. Which of the following is the power series representation of xe^{2x+1}?

$(A)\ x-2x^2+\left(2\right)^2\dfrac{x^3}{2!}-\left(2\right)^3\dfrac{x^4}{3!}+\cdots$

$(B)\ ex^2+2ex^4+\left(2\right)^2 e\dfrac{x^6}{2!}+\left(2\right)^3 e\dfrac{x^8}{3!}+\cdots$

$(C)\ ex+2ex^2+\left(2\right)^2 e\dfrac{x^3}{2!}+\left(2\right)^3 e\dfrac{x^4}{3!}+\cdots$

$(D)\ x+\left(2x+1\right)x+\dfrac{\left(2x+1\right)^2}{2!}x+\dfrac{\left(2x+1\right)^3}{3!}x+\cdots$

Answer_____

17. $\dfrac{d}{dx}\displaystyle\int_{4x}^{x^4}\left(t^4+2\right)^{11}dt =$

(A) $4x^3\left(\left(x^4\right)^4+2\right)^{11}-4\left(\left(4x\right)^4+2\right)^{11}$

(B) $\left(\left(x^4\right)^4+2\right)^{11}-\left(\left(4x\right)^4+2\right)^{11}$

(C) $4x^3\left(\left(x^4\right)^4+2\right)^{11}$

(D) $x^4\left(\left(x^4\right)^4+2\right)^{11}-4x\left(\left(4x\right)^4+2\right)^{11}$

Answer_____

18. What is the area of the region bounded by the curve $x = y^2 - 4$ and the y-axis?

(A) $\dfrac{28}{3}$

(B) 10

(C) $\dfrac{32}{3}$

(D) $\dfrac{34}{3}$

Answer _____

19. A particle moves horizontally with the velocity $v = \cos^2 \pi t$ from time $t = 0$ to $t = 10$. What is the total distance the particle travels in this time?

$(A)\,2 \qquad (B)\,\dfrac{5}{2} \qquad (C)\,4 \qquad (D)\,5$

Answer_____

20. What is the derivative of function $y(x)$ if $\tan(2y) = xe^y$?

$(A)\,\dfrac{e^y}{2\sec^2(2y) - xe^y} \qquad (B)\,\dfrac{e^y + xe^y}{2\sec^2(2y)} \qquad (C)\,\dfrac{e^y}{\sec^2(2y) - xe^y} \qquad (D)\,\dfrac{e^y}{2\sec(2y)\tan(2y) - xe^y}$

Answer_____

21. If $f(x) = \log \sqrt[4]{(3x+5)^3}$, where $x > -\dfrac{5}{3}$, then $f'(x) =$

(A) $\dfrac{3}{4(3x+5)\ln 10}$

(B) $-\dfrac{3}{4(3x+5)\ln 10}$

(C) $\dfrac{9}{4(3x+5)\ln 10}$

(D) $\dfrac{9}{(3x+5)\ln 10}$

Answer_____

22. What are all the values of x for which the series

$$(x+1) - \frac{(x+1)^2}{2} + \frac{(x+1)^3}{3} - \ldots + \frac{(-1)^{n+1}(x+1)^n}{n} + \ldots \text{ converges?}$$

(A) $-2 < x \leq 0$

(B) All real numbers

(C) $-2 < x < 0$

(D) $-2 \leq x \leq 0$

Answer _____

23. If n is a positive integer, then $\lim\limits_{n\to\infty}\dfrac{1}{n^3}\left[1^2+2^2+...+n^2\right]$ can be expressed as

$(A)\displaystyle\int_0^1 x^2\,dx$ $(B)\displaystyle\int_0^1 \frac{1}{x}\,dx$ $(C)\displaystyle\int_0^1 \frac{1}{x^3}\,dx$ $(D)\displaystyle\int_0^1 x^3\,dx$

Answer_____

24. What is $f'(x)$ if $f(x)=4^{\arctan\left(2x^5\right)}$?

$(A)\ 4^{\arctan\left(2x^5\right)}\cdot \sec^2\left(2x^5\right)\cdot \ln 4\cdot 10x^4$

$(B)\ 4^{\arctan\left(2x^5\right)}\cdot \dfrac{1}{\left(1+\left(2x^5\right)^2\right)\ln 4}\cdot 10x^4$

$(C)\ 4^{\arctan\left(2x^5\right)-1}\cdot \dfrac{10x^4}{1+\left(2x^5\right)^2}$

$(D)\ 4^{\arctan\left(2x^5\right)}\cdot \ln 4\cdot \dfrac{1}{1+\left(2x^5\right)^2}\cdot 10x^4$

Answer _____

25. $\lim\limits_{x \to 0} \dfrac{\int_0^x \dfrac{\ln(1+t)}{t}\,dt}{x} =$

$(A)\ 0 \quad (B)\ \infty \quad (C)\ 1 \quad (D)\ \ln 2$

Answer_____

26. What are all the horizontal asymptotes for the graph of $f(x) = \dfrac{5x}{\sqrt{x^2+1}}$?

(A) Horizontal asymptote is $y = 0$

(B) Horizontal asymptote is $y = 5$

(C) Horizontal asymptote is $y = -5$

(D) Horizontal asymptotes are $y = 5$ and $y = -5$

Answer_____

27. Given the following, which of the functions $f, g,$ and h has at least one point of inflection?

I. $f(x) = 2\sin 2x + 2$ II. $g'(x) = 5x^3$ III. $h(x) = \dfrac{6}{x} - \dfrac{1}{2x^2}, \ x \neq 0$

(A) I only

(B) II only

(C) I and II only

(D) I and III only

_____ **Answer**_____

28. Shown on the right is the slope field for which differential equation?

$(A)\ \dfrac{dy}{dx} = \dfrac{y^2 - 2y}{x}$

$(B)\ \dfrac{dy}{dx} = \dfrac{3x^2 - 4}{x}$

$(C)\ \dfrac{dy}{dx} = \dfrac{3y^2 - 4}{4}$

$(D)\ \dfrac{dy}{dx} = \dfrac{y - 2}{2x}$

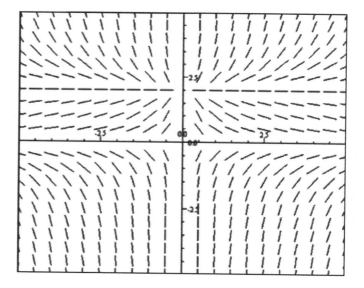

_____ **Answer** _____

29. A balloonist drops a sandbag from a balloon 160 feet above the ground. After t seconds, the sandbag is $160 - 16t^2$ feet above the ground. With what velocity does the sandbag strike the ground?

(A) $-32\sqrt{10}$

(B) $-16\sqrt{10}$

(C) $-24\sqrt{10}$

(D) -24

Answer_____

30. Let function $f(x)$ be positive while its first derivative $f'(x)$, and its second derivative $f''(x)$ all be negative on a closed interval $[a,b]$. The interval $[a,b]$ is partitioned into n equal length sub-intervals and these are used to compute an Upper Sum U, a lower sum L, and the Trapezoidal Rule Approximation T. If $I = \int_a^b f(x)dx$, which statement below is true?

(A) L < U < T < I

(B) L < T < I < U

(C) L < T < U < I

(D) L < I < T < U

Answer _____

Examination III

Section I Part B

Directions: Solve each of the following problems, using the space provided. Choose the best answer. Do not spend too much time on any one problem. A graphing calculator is required for some questions on this part of the exam.

In this test:

1. The exact numerical value of the correct answer does not always appear among the choices given. Then select from among the choices the number that best approximates the exact numerical value.

2. Unless otherwise specified, the domain of a function is assumed to be the set of all real numbers x for which $f(x)$ is a real number.

31. Let $f(x)$ be the function whose graph passes through the point $(0,1)$, and whose derivative is given by $f'(x) = \dfrac{1+e^{2x}}{2x+1}$. What is the approximation of $f(1)$ if Euler's method is used starting at $x = 0$ with the step size of 0.5?

$(A)\,2$ $(B)\dfrac{3+e}{4}$ $(C)\dfrac{9+e}{2}$ $(D)\dfrac{9+e}{4}$

Answer_____

32. As shown in the figure below the function $f(x)$ consists of a line segment from $(0,0)$ to $(3,5)$, another line segment from $(3,5)$ to $(7,5)$, and a quarter-circle with radius of 5. What approximately is the average value of this function on the interval $[0, 12]$?

$(A)\ 3.470$ $(B)\ 3.511$ $(C)\ 3.928$ $(D)\ 41.635$

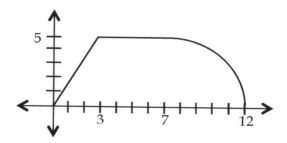

Answer_____

33. A particle moves along the x − axis so that at any time $t > 0$, its acceleration is given by $a(t) = \sin t - 3\sin t \cos t + 2t$. If the velocity of the particle is 6 m / sec at time $t = 3$ sec, then the velocity in m / sec of the particle at time $t = 7$ sec is approximately equal to

(A) 43.639 (B) 31.640 (C) 37.639 (D) 39.366

Answer_____

34. An object moves along the x − axis so that its position at any time t is given by $x(t) = -4\cos\dfrac{t}{2} + 3$. For which value of t is its speed the greatest on the interval $[4,8]$?

(A) $t = 4$

(B) $t = 5$

(C) $t = 6$

(D) $t = 7$

Answer _____

35. An object moving along a curve at time t has position $\left(x(t), y(t)\right)$ with $\dfrac{dx}{dt} = \cos\left(t^2\right)$

and $\dfrac{dy}{dt} = 2\sin\left(t^3\right)$ for $0 \le t \le 4$. At time $t = 2$ the object has position $\left(3, 4\right)$.

Which of the following is the position of the object at time $t = 4$?

$(A)\left(2.574, 2.613\right)$ $(B)\left(3.133, 3.972\right)$ $(C)\left(4.052, 4.012\right)$ $(D)\left(3.133, 4.711\right)$

Answer_____

36. If f is a continuous function defined by

$$f(x) = \begin{cases} x^2 - bx, & x \le 3 \\ 3\cos\left(\dfrac{\pi}{4}x\right), & x > 3 \end{cases}$$

then $b \approx$

$(A) -2.2929$ $(B) 3.7071$ $(C) -3.7071$ $(D) 2.2929$

Answer_____

Examination III

37. An outdoor thermometer registers a temperature of $100°F$. Twelve minutes after it is brought into a room where the temperature is $80°F$, the thermometer registers $90°F$. When will the thermometer register $85°F$?

(A) in 20 minutes (B) in 24 minutes (C) in 32 minutes (D) in 48 minutes

Answer_____

38. Which of the following is the equation of the line tangent to the curve described by the pair of parametric equations $x = 5t^3 - 5$ and $y = t^2 + 7$ at the point where $t = 2$?

(A) $15y - x = 130$

(B) $15y + x = 195$

(C) $10y + x = 145$

(D) $15y - 2x = 95$

Answer _____

39. The region bounded by the graphs of the equations $(x-4)^2 = y-2$ and $2y-3x+1=0$ is revolved around the x-axis. The volume of the solid generated is

(A) 162.90

(B) 218.36

(C) 342.99

(D) 436.71

Answer_____

40. A solid has its base the region bounded by the graph of $y = \dfrac{10}{1+x^2}$ and the horizontal line $y = 3$. What is the volume of the solid if every cross section by a plane perpendicular to the x-axis is a semicircle?

(A) 21.009 (B) 20.672 (C) 17.442 (D) 10.374

Answer _____

41.   A boy is standing on a dock watching a boat moving north away from him, at a speed of 5000ft / min. A girl is standing 1000 ft to the east of the boy and is watching the same boat. How fast is the boat moving away from the girl when it is 12500 ft away from the boy?

(A)538.280 (B)5023.949 (C)4984.08 (D)4321.957

Answer_____

42.   The region bounded by the graph of $y = -x^2 + 2$ and the $x-$axis is revolved around $y = 3$. What is the volume of the resulting solid?

(A)52.130 (B)12.506 (C)24.600 (D)27.842

Answer_____

43. The population $P(t)$ of a species satisfies the logistic differential equation $\dfrac{dP}{dt} = P\left(704 - \dfrac{P^2}{33}\right)$ where the initial population $P(0)=30$ and t is the time in years. For what value of P is the population growing the fastest?

(A) 88

(B) 64

(C) 97

(D) 121

Answer_____

44. Let $f(x)$ be a function that is differentiable for all x. The table below gives the value of $f'(x)$ for several values of x. If $f'(x)$ is always increasing, which statement about $f(x)$ must be true?

(A) $f(x)$ is an odd function.

(B) $f(x)$ is increasing on $[-4,4]$.

(C) $f(x)$ has a relative minimum at $x = 0$

(D) $f(x)$ has a relative maximum at $x = 0$.

x	-4	-2	0	2	4
$f'(x)$	-3	-1	0	1	3

Answer _____

45. A rectangle with vertices $(0,0)$, $(8,0)$, $(8,10)$ and $(0,10)$ is shown below. The graph of $y = x + \sqrt[3]{x}$ divides it into two regions. What is the probability that a point picked randomly from inside of this rectangle lies in the region above the graph?

(A) $\dfrac{15}{32}$ (B) $\dfrac{9}{20}$ (C) $\dfrac{7}{16}$ (D) $\dfrac{3}{8}$

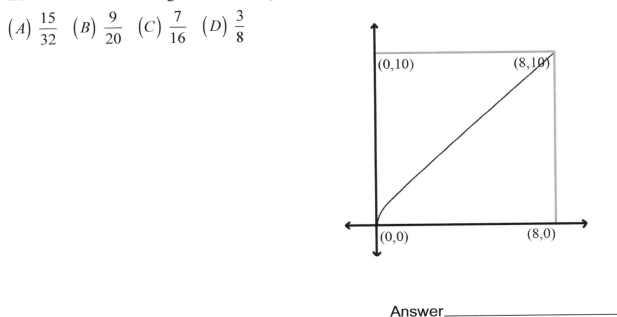

Answer_____

Section I Part A

Directions: Solve each of the following problems, using the space provided. Choose the best answer. Do not spend too much time on any one problem. Calculators may NOT be used on this part of the exam.

In this test: Unless otherwise specified, the domain of a function is assumed to be the set of all real numbers x for which $f(x)$ is a real number.

1. $\displaystyle\int_{3}^{\infty} \frac{1}{x^2\sqrt{x}}dx =$

(A) $\dfrac{\sqrt{3}}{3}$

(B) $\dfrac{2\sqrt{3}}{27}$

(C) $\dfrac{2\sqrt{3}}{9}$

(D) $\dfrac{2}{27}$

Answer _____

2. Let $f(x)$ be a continuous function. What is the value of $\displaystyle\int_{3}^{15} f(x)dx$ if it is given that

$\displaystyle\int_{1}^{5} f(3x)dx = 7$?

(A) $\dfrac{7}{3}$ (B) 7 (C) 14 (D) 21

Answer _____

Examination IV

3. $\dfrac{d}{dx}\displaystyle\int_0^{\frac{\pi}{4}}(\cos t - \sin t)\,dt$ is equal to which of the following?

(A) $\sqrt{2}+1$ (B) $\sqrt{2}$ (C) $\sqrt{2}-1$ (D) 0

Answer_____

4. $\displaystyle\lim_{h\to 0}\dfrac{9\left(\frac{1}{3}(x+h)\right)^3 - 9\left(\frac{1}{3}x\right)^3}{h} =$

(A) $\dfrac{x^2}{3}$

(B) 0

(C) $9x^2$

(D) x^2

Answer_____

5. Which of the following is true for the given series:

$$\sum a_n = \sum_{n=1}^{\infty} (-1)^n \left(\frac{2}{n+2} \right) ; \qquad \sum b_n = \sum_{n=1}^{\infty} (-1)^n \left(\frac{n^3 + 4}{5^n} \right)$$

(A) Both Conditionally Convergent

(B) Both Absolutely Convergent

(C) $\sum a_n$ is Conditionally Convergent, $\sum b_n$ is Absolutely Convergent

(D) $\sum a_n$ is Conditionally Convergent, $\sum b_n$ is Divergent

Answer_____

6. $\int \sin^{-1} x \, dx =$

$(A)\ x \sin^{-1} x + \sqrt{1 - x^2} + C \qquad (B)\ x \sin^{-1} x - \sqrt{1 - x^2} + C$

$(C)\ \dfrac{x^2}{2} \sin^{-1} x - \dfrac{\sqrt{1 - x^2}}{2} + C \qquad (D)\ x \sin^{-1} x + \dfrac{\sqrt{x^2 - 1}}{2} + C$

Answer_____

7. $\lim\limits_{x\to 2}\dfrac{\sqrt{x^2+5}-3}{x^2-4}$ is

(A) nonexistent (B) 0 (C) $\dfrac{1}{6}$ (D) $\dfrac{1}{3}$

Answer_____

8. The graph of $f'(x)$ on the interval $[-1,4]$ is shown below. On what intervals is the graph of $f(x)$ concave up?

(A) $(0.5,3.5)$

(B) $(-1,0.5)\cup(3.5,4)$

(C) $(2,4)$

(D) $(0.5,4)$

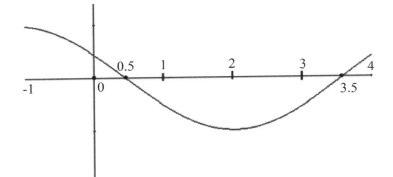

Answer_____

Examination IV

9. 2000 feet of wire will be used to construct six identical rectangular enclosures as shown on the diagram below. What are the dimensions (in feet) that maximize the enclosed area of the large rectangle ?

$(A)\ l=400, w=200$

$(B)\ l=500, w=125$

$(C)\ l=\dfrac{1000}{3}, w=250$

$(D)\ l=\dfrac{500}{3}, w=375$

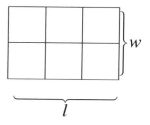

Answer_____

10. $\displaystyle\int_{m}^{m+h} f(x)\ dx - \int_{m}^{h} f(x)\ dx =$

$(A)\displaystyle\int_{0}^{h} f(x)\ dx$

$(B)\displaystyle\int_{h}^{h+m} f(x)\ dx$

$(C)\displaystyle\int_{h}^{m} f(x)\ dx$

$(D)\displaystyle\int_{m}^{m+h} f(x)\ dx$

Answer_____

11. An equation of the line tangent to the curve $x^2 + y^2 = 25$ at the point $(-4, 3)$ is

(A) $3y - 4x = 25$ $\qquad$ (B) $4y - 3x = 25$ $\qquad$ (C) $3y + 4x = -25$ $\qquad$ (D) $3y - 4x = 20$

Answer _____

12. What is the area of the region bounded by the graphs of $y = x^2 - 4$ and $y = 3x$?

(A) $\dfrac{56}{3}$

(B) $\dfrac{45}{6}$

(C) 21

(D) $\dfrac{125}{6}$

Answer _____

13. What are all values of x for which the series $\sum_{n=1}^{\infty} \dfrac{(2x-1)^n}{\sqrt[3]{n^2}}$ converge?

$(A)\ 0 < x < 1$

$(B)\ 0 \le x < 1$

$(C)\ 0 < x \le 1$

$(D)\ -1 \le x < 1$

Answer_____

14. Which of the following is a power series representation for $f(x) = \dfrac{1}{4+6x}$ if $|x| < \dfrac{2}{3}$?

$(A)\ \dfrac{1}{4}\sum_{n=0}^{\infty}(-1)^n\left(\dfrac{3}{2}\right)^n x^n$

$(B)\ \dfrac{1}{6}\sum_{n=0}^{\infty}(-1)^n\left(\dfrac{2}{3}\right)^n x^n$

$(C)\ \dfrac{1}{4}\sum_{n=0}^{\infty}(-1)^{2n}\left(\dfrac{3}{4}\right)^n x^n$

$(D)\ \dfrac{1}{4}\sum_{n=0}^{\infty}\left(\dfrac{3}{2}\right)^n x^n$

Answer_____

15. Shown below is the graph of $f(x)$. Which of the following limits exist?

I. $\lim\limits_{x \to 0} f(x)$

II. $\lim\limits_{x \to 1} f(x)$

III. $\lim\limits_{x \to 2} f(x)$

(A) I only $\quad (B)$ I and II only $\quad (C)$ I and III only $\quad (D)$ I, II, and III

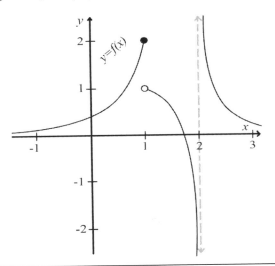

Answer _____

t	0	1	2	3	4	5
$f(t)$	0	50	90	200	275	350

16. The table above shows the total distance $f(t)$ a bicyclist has traveled along a straight road in meters during t seconds. What is approximately the velocity of the bicyclist at $t = 4$ seconds?

(A) 75 m/sec $\quad (B)$ 50 m/sec $\quad (C)$ 40 m/sec $\quad (D)$ 25 m/sec

Answer _____

17. What is the derivative of $\dfrac{4x^2 - 3x + 7}{5x}$?

$(A)\ \dfrac{4x^2 + 7}{25x^2}$

$(B)\ \dfrac{4x^2 - 7}{5x^2}$

$(C)\ \dfrac{7 - 4x^2}{5x^2}$

$(D)\ \dfrac{8x - 3}{5}$

Answer_____

18. What is the area of one "petal" of the curve $r = 2\sin 5\theta$?

$(A)\ \dfrac{2}{5}$ $(B)\ \dfrac{4}{5}$ $(C)\ \dfrac{\pi}{10}$ $(D)\ \dfrac{\pi}{5}$

Answer_____

19. If $f(x) = \sqrt{6\sin x + 9}$, then the derivative of f at $x = 0$ is

$(A)\dfrac{1}{2\sqrt{3}}$ $(B)0$ $(C)1$ $(D)\sqrt{3}$

Answer _____

20. Which of the following statements are true regarding the series below:

$$\frac{1}{2} + \frac{(x-25)}{4} + \frac{(x-25)^2}{8} + \ldots + \frac{(x-25)^n}{2^{n+1}} + \ldots?$$

I. Interval where the series converges is $23 < x < 27$

II. The sum of the series where it converges is equal to $\dfrac{1}{27-x}$

III. If $x = 28$, series converges to $\dfrac{1}{3}$

IV. Interval where the series converges is $21 < x < 29$

(A) I only (B) III only (C) I and II only (D) I and IV only

Answer _____

21. Given the equation of the curve $xy = 5 + y$, where y is the twice differentiable function of x, what is y''?

(A) $\dfrac{1-x+y}{(x-1)^2}$

(B) $\dfrac{2y}{(x-1)^2}$

(C) $-\dfrac{2y}{(x-1)^2}$

(D) $\dfrac{-y}{(x-1)}$

Answer_____

22. A 25 feet ladder leans against a vertical wall. The foot of the ladder is pulled away from the wall horisontally at the rate of 6 ft / sec. What is the rate (in ft/sec) at which the top of the ladder is sliding down the wall when the top of the ladder is 15 feet above the ground?

(A) -3

(B) -5

(C) -7

(D) -8

Answer _____

23. Which of the following are the absolute extrema of the function $f(x) = 3x^2 - 9x + 8$ on the interval $[-2,3]$?

(A) max $= 8$, min $= \dfrac{5}{4}$

(B) max $= 38$, min $= 8$

(C) max $= 38$, min $= \dfrac{5}{4}$

(D) max $= 9$, min $= 0$

Answer _____

24. Shown below is the graph of the differentiable function $f(x)$. Which of the following statements about $f(x)$ must be true?

I. $f'(c) = 0$ for some value c in the closed interval $[0,5]$

II. $f(c) = 0$ for some value c in the open interval $(0,5)$

III. $f'(x)$ is an odd function

(A) I only (B) II only (C) I and II only (D) II and III only

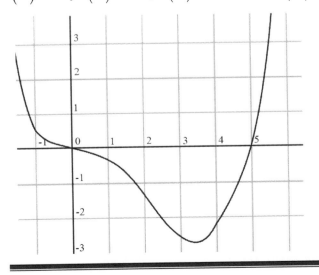

Answer _____

25. How many local extrema does the function $f(x) = (x+1)^2(x-2)^3$ have on the open interval $(-9, 5)$?

(A) One

(B) Two

(C) Three

(D) Four

Answer_____

26. A particle moves to the right along the x-axis until it reaches the origin and then moves along the x-axis to the left. The velocity of this particle is given by $v(t) = 20 - 4t$ for $t \geq 0$. What is the position of the particle at any time t?

(A) $-2t^2 + 20t - 50$

(B) $-2t^2 + 20t + 50$

(C) $-4t^2 + 20t$

(D) $-4t^2 + 20t - 50$

Answer_____

27. Let $f(t) = \dfrac{3}{t^2}$. For what value of t is $f'(t)$ equal to the average rate of change of f on a closed interval $[a,b]$?

$(A) \sqrt[3]{\dfrac{2a^2 b^2}{a-b}}$ $(B) \sqrt[3]{\dfrac{2a^2 b^2}{a+b}}$ $(C) -\sqrt[3]{\dfrac{2a^2 b^2}{a+b}}$ $(D) \sqrt[3]{\dfrac{a^2 b^2}{a+b}}$

Answer _____

28. Shown below is the slope field of which differential equation?

$(A)\ \dfrac{dy}{dx} = xy - y$

$(B)\ \dfrac{dy}{dx} = 3y^2 - 4$

$(C)\ \dfrac{dy}{dx} = xy$

$(D)\ \dfrac{dy}{dx} = 5x$

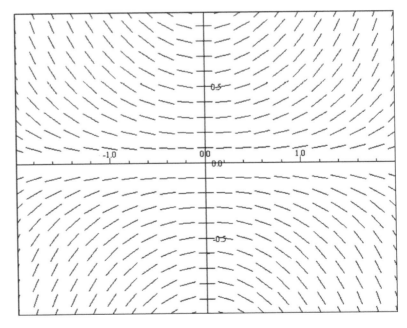

Answer _____

29. What is equation of the tangent line to the graph of $y = x - e^{-x}$ that is parallel to the line $6x - 2y = 15$?

$(A)\, y = 3x + 2\ln 2 - 2$ $(B)\, y = 3x + 2\ln 2 + 2$ $(C)\, y = 3x - 2\ln 2 - 2$ $(D)\, y = 3x - 2\ln 2 + 2$

Answer_____

30. The graph of the function $f(x)$ is shown below. If $g(x) = \int_x^1 f(t)\,dt$, what is the limit of $g'(x)$ as x approaches 4?

$(A)\, 6$ $(B)\, 3$ $(C)\, -2$ $(D)\, -3$

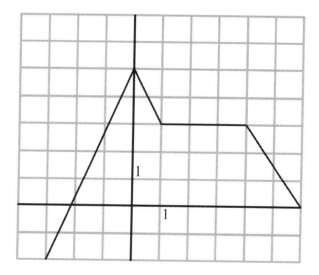

Answer_____

Examination IV

Directions: Solve each of the following problems, using the space provided. Choose the best answer. Do not spend too much time on any one problem. A graphing calculator is required for some questions on this part of the exam.

In this test:

1. The exact numerical value of the correct answer does not always appear among the choices given. Then select from among the choices the number that best approximates the exact numerical value.

2. Unless otherwise specified, the domain of a function is assumed to be the set of all real numbers x for which $f(x)$ is a real number.

31. What approximately is the area under the curve $f(x) = x^2 + 4x + 6$ on the interval $[2,6]$ using the right-hand Riemann sum where P is the partition of $[2,6]$ determined by $\{2,4,5,6\}$?

(A) 193

(B) 160

(C) 195

(D) 143

Answer _____

32. The second derivative of function f is given by $f''(x) = \sin(e^{0.3x}) + \dfrac{x}{35}$. How many relative maximum points does $f'(x)$ have in the interval $0 < x < 10$?

(A) One

(B) Two

(C) Three

(D) More than three

Answer _____

Examination IV

33. Given $f''(x) = 3 + 4\cos x$, $f'(0) = 0$, and $f(0) = 0$. The line, tangent to the graph of $f(x)$ and parallel to the segment connecting the endpoints of the interval $[0,5]$, touches $f(x)$ at $x = ?$

$(A)0.596$ $(B)1.018$ $(C)1.381$ $(D)2.073$

Answer_____

34. Given differentiable function $f(x) = 2 + \int_{0}^{2x} \sin(t^3) dt$. Which of the following is the smallest positive number c for which $f'(c) = 0$?

$(A)\ 0.327$

$(B)\ 0.463$

$(C)\ 0.656$

$(D)\ 0.732$

Answer _____

35. Let R be the region bounded by the graphs: $y = 2x$, $y = kx$, $x = 0$, $x = 10$

If the area of R $= 50$ and $k < 2$, the value of k is :

(A) 1.00

(B) 1.25

(C) 1.50

(D) 1.75

Answer _____

36. Which of the following series diverge?

I. $\sum\limits_{n=1}^{\infty} \dfrac{3n^{\frac{9}{2}}(8n+1)!}{4n^6(8n)!}$ II. $\sum\limits_{n=1}^{\infty} \dfrac{1}{n^{\frac{1}{3}}}$ III. $\sum\limits_{n=1}^{\infty} \dfrac{1}{e^n}$

(A) I only (B) II only (C) I and II only (D) II and III only

Answer _____

37. Let $f(x)$ be a continuous and differentiable function on the interval $0 \leq x \leq 1$, and let $g(x) = f(2x)$. The table below gives values of $f'(x)$, the derivative of $f(x)$. What is the value of $g'(0.2)$?

(A) 2.32

(B) 3.01

(C) 4.64

(D) 6.02

x	0.1	0.2	0.3	0.4	0.5	0.6
f ' (x)	2.05	2.32	2.56	3.01	3.52	3.75

Answer_____

38. The velocity of a particle is given by $v(t) = t \sin t^2$. What is the distance traveled by the particle from time $t = 0$ to time $t = \dfrac{\pi}{3}$?

(A) 0.225

(B) 0.235

(C) 0.272

(D) 0.342

Answer_____

39. The equation $P(t) = 5500 - 3400e^{-0.089t}$ describes the growth of a population of a species, where $t \geq 0$ is the number of years. This population will approach a limiting value as time goes on. During which year will the population reach half of this limiting value?

(A) First

(B) Second

(C) Third

(D) Fourth

Answer _____

40. What is the volume of the solid generated when the region in the first quadrant bounded by the graphs of $y = x^3$ and $y = 4x$ is revolved around the line $x = 5$?

(A) 12.566

(B) 38.204

(C) 73.723

(D) 98.855

Answer _____

41. The acceleration of a particle moving along the line is given by $a(t) = t\cos(t^2)$. If at time $t = 0\sec$, its velocity is $2m/\sec$ and position is $4m$, what is the position of the particle at time $t = 7\sec$?

(A) 14.303 m

(B) 4.303 m

(C) 17.697 m

(D) 18.303 m

Answer_____

42. A wire 100 inches long is to be cut into two pieces. One of the pieces will be bent into a circle and the other into a square. What is the minimum sum of the areas of the circle and the square?

(A) 350.06 in^2 $\quad(B)$ 625 in^2 $\quad(C)$ 331.07 in^2 $\quad(D)$ 427.81 in^2

Answer _____

43. Let the function $f(x) = \sin\left(4x + \dfrac{\pi}{6}\right)$, and let $P(x)$ be the third-degree Taylor polynomial

for f about $x = 0$. What is the maximum value of $\left| f\left(\dfrac{1}{2}\right) - P\left(\dfrac{1}{2}\right) \right|$?

(A) 0.333

(B) 0.667

(C) 0.783

(D) 0.823

Answer _____

---

44. A solid has its base the region R bounded by the graph of $y = 3\sqrt{x}$, the horizontal line $y = 9$, and the y-axis. Which of the following is the integral expression that gives the volume of the solid if every cross section perpendicular to the y-axis is a rectangle whose height is 5 times the length of its base in region R?

$(A)\ 5\displaystyle\int_0^9 \left(3\sqrt{x}\right)^2 dx$ $(B)\ \displaystyle\int_0^9 \left(3\sqrt{x}\right)\left(15\sqrt{x}\right)\, dx$ $(C)\ \dfrac{5}{81}\displaystyle\int_0^9 y^4 dy$ $(D)\ \dfrac{5}{9}\displaystyle\int_0^9 y^2 dy$

Answer _____

45. Let f be a function having derivatives of all orders for all real numbers x. The third-degree Taylor polynomial for f about $x = 4$ is given by $T(x) = \dfrac{1}{9} + 5(x-4)^2 - 8(x-4)^3$.

If $\left| f^{(4)}(x) \right| \leq \dfrac{1}{4}$ for $3.5 \leq x \leq 4$, order the following from greatest to least.

I. Maximum value of $\left| T(x) - f(x) \right|$ for $3.5 \leq x \leq 4$ II. $\left| f''(4) \right|$ III. $\left| f(4) \right|$

(A) I, II, III

(B) II, I, III

(C) II, III, I

(D) III, II, I

Answer_____

Examination V

Section I Part A

Directions: Solve each of the following problems, using the space provided. Choose the best answer. Do not spend too much time on any one problem. Calculators may NOT be used on this part of the exam.

In this test: Unless otherwise specified, the domain of a function is assumed to be the set of all real numbers x for which $f(x)$ is a real number.

1. $\lim\limits_{x \to 0} \dfrac{3 - \sqrt{9+x}}{x}$ is

(A) nonexistent

(B) 6

$(C) -\dfrac{1}{9}$

$(D) -\dfrac{1}{6}$

Answer _____

2. A function f satisfies the given conditions:

$$\lim_{x \to \infty} f(x) = 0 \qquad \lim_{x \to -\infty} f(x) = 0 \qquad \lim_{x \to 0^+} f(x) = \infty \qquad \lim_{x \to 0^-} f(x) = \infty$$

Which of the following is a possible graph for f, assuming that it does not cross a horizontal asymptote?

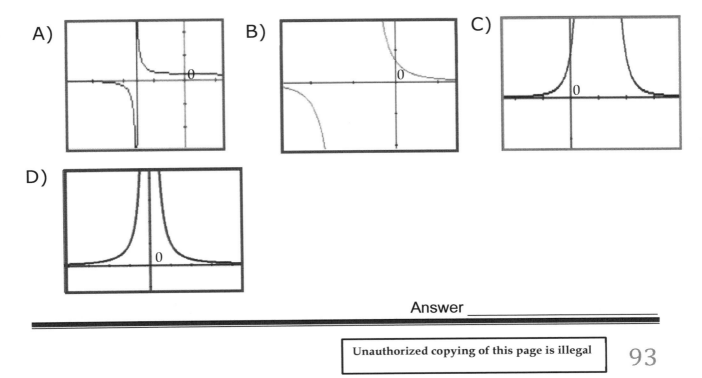

A)

B)

C)

D)

Answer _____

3. Let $y = f(x)$ be a particular solution to the differential equation $\dfrac{dy}{dx} = xy^3$ with $f(1) = 2$. What is the approximate value of $f(1.1)$ if the equation of the line tangent to the graph of $y = f(x)$ at $x = 1$ is being used to approximate $f(1.1)$?

(A) 3

(B) 2.2

(C) 2.6

(D) 2.8

Answer_____

4. Given the series $\displaystyle\sum_{n=1}^{\infty} a_n = \sum_{n=1}^{\infty} \dfrac{6n(n^2+4)(n^3-3n+2)}{e^n(n^2+4n)(n^4-1)}$. When using the Limit Comparison Test, what choice of $\displaystyle\sum_{n=1}^{\infty} b_n$ would be appropriate to prove the convergence of $\displaystyle\sum_{n=1}^{\infty} a_n$?

(A) $\displaystyle\sum_{n=1}^{\infty} \dfrac{1}{n^2}$ (B) $\displaystyle\sum_{n=1}^{\infty} \dfrac{1}{n^3}$ (C) $\displaystyle\sum_{n=1}^{\infty} \dfrac{1}{ne^n}$ (D) $\displaystyle\sum_{n=1}^{\infty} \dfrac{1}{e^n}$

Answer_____

5. Which of the following series converge?

I. $\displaystyle\sum_{n=2}^{\infty}\frac{1}{n(\ln n)^4}$ II. $\displaystyle\sum_{n=1}^{\infty}\frac{3+\sin n}{n^4}$ III. $\displaystyle\sum_{n=1}^{\infty}\frac{7n^2-5}{e^n(n+3)^2}$

(A) I and II only

(B) II and III only

(C) I and III only

(D) I, II, and III

Answer _____

6. What is y'' if $\sin y = y + 5x$?

$(A)\ \dfrac{5}{-1+\cos y}$ $(B)\dfrac{5\sin y}{\left(-1+\cos y\right)^2}$ $(C)-\dfrac{25\cos y}{\left(1+\cos y\right)^2}$ $(D)\dfrac{25\sin y}{\left(-1+\cos y\right)^3}$

Answer _____

7. Graph of $f''(x)$, the second derivative of function $f(x)$ is shown on the right. What must be true about $f(x)$?

I. $f(x)$ is concave downward for all $a < x < b$.

II. $f(x)$ must have a critical points at $x = a$ and $x = c$.

III. The slope of $f(x)$ increases when $x < a$.

(A) I only $\quad (B)$ III only $\quad (C)$ I and III only $\quad (D)$ I,II and III

Answer_____

8. At which value(s) of x does the function $f(x) = (x+1)^3 (x-6)^2$ have a local minimum?

(A) 6 only

(B) 6 and -1 only

(C) $\dfrac{16}{5}$ and -1 only

(D) 6 and $\dfrac{16}{5}$ only

Answer_____

9. The curves $f(x)$, $f'(x)$, and $f''(x)$ are shown.

A function $f(x)$ is increasing and concave down.

What are the correct labels for graphs I, II, and III?

$$\begin{array}{cccc} & \text{I} & \text{II} & \text{III} \end{array}$$

(A) $f'(x)$ $f(x)$ $f''(x)$

(B) $f''(x)$ $f'(x)$ $f(x)$

(C) $f(x)$ $f''(x)$ $f'(x)$

(D) $f(x)$ $f'(x)$ $f''(x)$

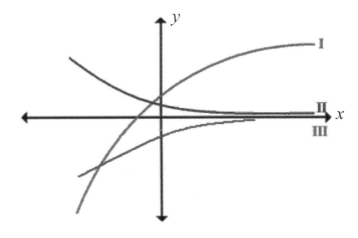

Answer _____

10. $\lim\limits_{x \to 0} \dfrac{e^x + 2e^{-x} - 3}{x - \sin 2x} =$

(A) -3

(B) -1

(C) 1

(D) 3

Answer _____

11. What is the area under the curve of $F(x) = f(x) - g(x)$ for $-3 \le x \le 3$ if $f(x) = 12 - g(x)$?

$(A)\ 72 - 2\int_{-3}^{3} g(x)dx$

$(B)\ 72 + \int_{-3}^{3} f(x)dx$

$(C)\ 72 + 2\int_{-3}^{3} g(x)dx$

$(D)\ \int_{-3}^{3} (6 - g(x))dx$

Answer _____

12. The graph of the velocity $v(t)$ of a particle is shown below. What is the total distance traveled by the particle from time $t = 0$ to $t = 5$?

$(A)\ 2\pi - 1.5 \quad (B)\ 4\pi + 1.5 \quad (C)\ 2\pi + 1.5 \quad (D)\ 2\pi + \sqrt{10}$

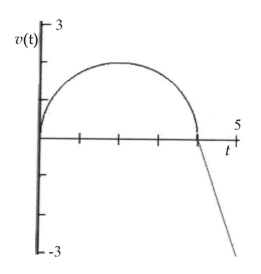

Answer _____

13. The region bounded by the graphs of $y = -\sin x$, $y = \sin x$, $x = 0$, and $x = \pi$ is revolved around $y = 2$. What is the volume of the resulting solid?

$(A)\,8\pi$ $(B)\,16\pi$ $(C)\,4\pi$ $(D)\,32\pi$

Answer _____

14. Which of the following is a power series representation for $\ln(1+2x)$?

$(A)\ \ 2x - \dfrac{4x^2}{2} + \dfrac{8x^3}{3} - \dfrac{16x^4}{4} + \dfrac{32x^5}{5} + ...$

$(B)\ \ 2 - 4x + 8x^2 - 16x^3 + 32x^4 + ...$

$(C)\ \ 2x^2 - 4x^4 + 8x^6 - 16x^8 + 32x^{10} + ...$

$(D)\ \ 2x^2 - \dfrac{4x^4}{2} + \dfrac{8x^6}{3} - \dfrac{16x^8}{4} + \dfrac{32x^{10}}{5} + ...$

Answer _____

15. Which of the following is equal to the area of the region bounded by $y = (x-1)^2$ and $y = x + 1$?

$(A)\ \dfrac{9}{2}$ $(B)\ 3$ $(C)\ 18$ $(D)\ \dfrac{11}{2}$

Answer_____

16. Let $P(x) = 7x + \dfrac{5}{2}x^2$ be the second-degree Taylor polynomial for $f(x)$ around $x = 0$. If $g(x)$ is the inverse of $f(x)$, what is $g'(0)$?

$(A)\ 1$ $(B)\ \dfrac{1}{5}$ $(C)\ \dfrac{1}{7}$ $(D)\ 5$

Answer _____

17. $\int \sec^2 \dfrac{x}{9} \tan \dfrac{x}{9}\, dx =$

$(A)\ \dfrac{1}{3}\tan^2 \dfrac{x}{9} + C$

$(B)\ 9\tan^2 \dfrac{x}{9} + C$

$(C)\ \dfrac{9}{2}\tan^2 \dfrac{x}{9} + C$

$(D)\ \dfrac{9}{2}\tan^3 \dfrac{x}{9} + C$

Answer _____

18. A particle moves along the x-axis so that its velocity at any time $t \geq 0$ is given by $v(t) = \pi \sin(\pi t)$. What is the total distance the particle travels from $t = 0$ to $t = 2$?

$(A)\ 0 \qquad (B)\ 2 \qquad (C)\ 2\pi \qquad (D)\ 4$

Answer _____

19. What is the the volume of the solid generated when the region in the first quadrant bounded by the graph of the function $y = 3x - x^2$ is revolved around the $x - $axis?

$(A)\dfrac{9\pi}{2}$

$(B)\dfrac{81\pi}{10}$

$(C)\dfrac{32\pi}{5}$

$(D)\dfrac{81}{10}$

Answer_____

x	-1	2	4	6	7
f (x)	3	5	-2	8	-6
f'(x)	2	6	9	7	5

20. Let $g(x)$ be the inverse of the function $f(x)$. Given the following values on the table above, at which value $x = a$ will $g'(a) = \dfrac{1}{6}$?

$(A) -1$

$(B)\ 2$

$(C)\ 3$

$(D)\ 5$

Answer_____

Examination V

21. Which of the following is the solution to the differential equation $\dfrac{dy}{dx} = \dfrac{x-5}{y+1}$; $y \neq -1$, subject to initial conditions $y(0) = 3$?

$(A)\, y = x - 6$

$(B)\, y = \sqrt{(x-5)^2 - 9} - 1$

$(C)\, y = \sqrt{(x-5)^2 - 9}$

$(D)\, y = \sqrt{(x-5)^2 - 16}$

Answer_____

22. If $\displaystyle\int_{60}^{150} v(t)\,dt = 127$, and $\displaystyle\int_{80}^{150} v(t)\,dt = -78$, what is $\displaystyle\int_{60}^{80} v(t)\,dt$?

$(A)\ 49 \qquad (B)\ 205 \qquad (C)\ -205 \qquad (D)\ -49$

Answer_____

Examination V

23. A projectile is fired from 16 ft above the ground with an initial velocity of 10 ft/s. Its downward acceleration is represented by $a(t) = -12t^3$. Which of the following represents the height of the projectile at time t, if time $t = 0$ when it is launched?

$(A) -\dfrac{4}{5}t^5 + 10t + 16$

$(B) -\dfrac{3}{5}t^5 + 10t + 16$

$(C) -3t^4 + 10$

$(D) -\dfrac{3}{5}t^5 - 10t + 16$

Answer _____

24. What is the length of the curve of $f(x) = \ln(\sec x)$ between $x = 0$ and $x = \dfrac{\pi}{4}$?

$(A)\ \ln\left(\sqrt{2}+1\right)$

$(B)\ \ln\left(\dfrac{\sqrt{2}}{2}+1\right)$

$(C)\ 1-\dfrac{\sqrt{2}}{2}$

$(D)\ 1+\dfrac{\sqrt{2}}{2}$

Answer _____

Examination V

25. Shown below is the slope field for which differential equation?

(A) $\dfrac{dy}{dx} = 1 + y^2$

(B) $\dfrac{dy}{dx} = x - y$

(C) $\dfrac{dy}{dx} = 1 + x^2$

(D) $\dfrac{dy}{dx} = 1 - y^2 + x^2$

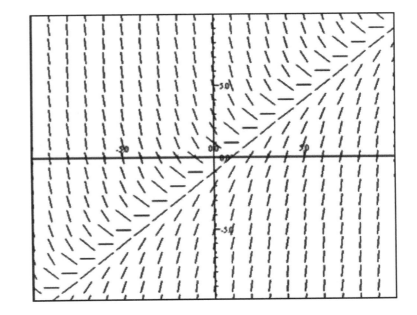

Answer_____

---

26. An electron travels along the path defined by the polar equation $r = 4 - 3\cos\theta$ and has the position $\left(x(t), y(t)\right)$ at time t. Given that $\dfrac{d^2 y}{d\theta^2} = \dfrac{dy}{dt}$ and $\theta = 0$ when $t = 0$, what is $\dfrac{dy}{dt}$ at $\theta = \dfrac{\pi}{4}$?

(A) $-6 - 2\sqrt{2}$ (B) $\dfrac{3\sqrt{2}}{2}$ (C) $6 - 2\sqrt{2}$ (D) $6 + 2\sqrt{2}$

Answer _____

27. The base of a solid is the region in the first quadrant bounded by the line $x + y = 4$ and the coordinate axes. What is the volume of the solid if every cross section perpendicular to the x-axis is a semicircle?

$(A) \dfrac{4\pi}{3}$ $\qquad$ $(B) \dfrac{8\pi}{3}$ $\qquad$ $(C) \dfrac{32\pi}{3}$ $\qquad$ $(D) \dfrac{64\pi}{3}$

Answer _____

28. The length of a curve $y = f(x)$ from $x = a$ to $x = b$ is given by $\displaystyle\int_a^b \sqrt{9x^2 + 12x + 5}\, dx$. Which of the following could $f(x)$ be equal to?

$(A)\ 3x + 2$ $\quad$ $(B)\ \dfrac{3x^2}{2} + 5x + C$ $\quad$ $(C)\ \dfrac{3x^2}{2} + 2x + C$ $\quad$ $(D)\ 3x^3 + 6x^2 + 5x + C$

Answer _____

29. What is the volume of the solid generated when the region bounded by the graphs of $x = \sqrt{y}$ and $x = \dfrac{1}{3}y$ is rotated around the y-axis?

$(A)\dfrac{9\pi}{2}$ $\quad$ $(B)9\pi$ $\quad$ $(C)\dfrac{27\pi}{2}$ $\quad$ $(D)\,18\pi$

Answer_____

30. The coefficient of $(x-3)^2$ in the Taylor series for $f(x) = \arctan x$ about $x = 3$ is:

$(A)\dfrac{3}{100}$ $\quad$ $(B)-\dfrac{3}{100}$ $\quad$ $(C)-\dfrac{3}{50}$ $\quad$ $(D)\dfrac{3}{50}$

Answer_____

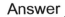

Examination V

Directions: Solve each of the following problems, using the space provided. Choose the best answer. Do not spend too much time on any one problem. A graphing calculator is required for some questions on this part of the exam.

In this test:

1. The exact numerical value of the correct answer does not always appear among the choices given. Then select from among the choices the number that best approximates the exact numerical value.

2. Unless otherwise specified, the domain of a function is assumed to be the set of all real numbers x for which $f(x)$ is a real number.

31. For the curve given parametrically by $x = \sin(5t), y = \cos(5t)$, where $2.4 \le t \le 3$, what is the point on the curve at which the tangent line to this curve is vertical?

$(A)\left(\dfrac{\sqrt{2}}{2}, 0\right)$ $\quad$ $(B)\ (1, 0)$ $\quad$ $(C)\ (1, 1)$ $\quad$ $(D)\left(\dfrac{1}{2}, 0\right)$

Answer_____

32. While on a road trip, a girl drives for 8 hours. Her velocity, recorded at random times in miles per hour, is given in the table below. Based on the data in the table, how many miles has she driven after 8 hours?

(A) 320 miles $\quad$ (B) 325 miles $\quad$ (C) 328 miles $\quad$ (D) 330 miles

t	0	1	5	6	8
$v(t)$	25	30	60	45	20

Answer_____

Examination V

33. At 3:00 PM car A is 50 miles south of car B and is driving north at a rate of 25mph. If car B is driving west at a rate of 15 mph, at what time is the distance between the cars minimal?

(A) 4:12 PM (B) 4:28 PM (C) 4:45 PM (D) 3:44 PM

Answer_____

34. What is the average value of the function $f(x) = 150\sin\left(\dfrac{2\pi}{365}(x-80)\right) + 670$ on the

interval $[1,181]$?

(A) 688.22

(B) 714.87

(C) 345.44

(D) 1439.62

Answer _____

35. The number of a certain bacteria doubles every 15 sec. Its rate of growth is directly proportional to the number of bacteria present. If the initial number of bacteria is 75,000, what is the number of bacteria after 15 min?

(A) 6.485×10^{27} (B) 9.067×10^{28} (C) 8.647×10^{22} (D) 6.452×10^{15}

Answer_____

36. Cream puffs are being filled with custard at the rate $r(t)=80t^{\frac{1}{4}}$ cream puffs per minute, with $t\geq0$ measured in minutes. If there were 700 filled cream puffs to begin with, how many cream puffs are filled after 625 minutes?

$(A)100700$ $(B)200000$ $(C)200700$ $(D)300000$

Answer _____

37. Let $f(t) = \arcsin(t)$ for $0 \le t \le 1$. For what value of t is the instantaneous rate of change of $f(t)$ equal to the average rate of change of $f(t)$ on the closed interval $[0.5, 1]$?

(A) 0.4726 (B) 0.8787 (C) 0.6868 (D) 0.9435

Answer_____

38. What is the length of curve given parametrically: $x = \sin(t^2)$, $y = 2\cos(\ln t)$, $\dfrac{\pi}{2} < t < \dfrac{3\pi}{2}$?

(A)1.947

(B)6.35

(C)13.063

(D)12.499

Answer_____

39. Which of the following is the best approximation of the area of the region between $y = \ln x$ and $y = \log x$ from their point of intersection to $x = 4$?

(A) 1.326 (B) 1.440 (C) 1.513 (D) 1.682

Answer_____

40. Consider the area of the region bounded by the graph of $y = \left(x - \dfrac{1}{2} \right)^2 + \dfrac{1}{2}$ and the x-axis in the interval $[0,2]$. Arrange the following in increasing order:

(I) Area calculated using Right Riemann Sums with four subintervals

(II) Area calculated using Left Riemann Sums with four subintervals

(III) Area calculated by evaluating the integral.

(A) I, III, II

(B) I, II, III

(C) II, I, III

(D) II, III, I

Answer _____

41. The position $r(t)$ of a particle at time t is given by the vector $\left(3\cos(4t),-\sin^2 t\right)$. What is the particle's velocity at $t = 3$?

$(A)(6.439,0.279)$ $(B)(6.439,-0.279)$ $(C)(-1.609,0.279)$ $(D)(-6.439,-0.279)$

Answer_____

42. The base of a container is defined by the region bounded by the graph of $y = -x^2 + 6x + 9$ in the first quadrant. The cross-sections of the container are semicircles perpendicular to the x-axis. Which of the following represents the volume of the container?

$(A)2\pi \int\limits_{0}^{7.24264} \left(\frac{1}{2}\left(-x^2+6x+9\right)\right)^2 dx$

$(B)\dfrac{\pi}{2} \int\limits_{0}^{7.24264} \left(\frac{1}{2}\left(-x^2+6x+9\right)\right)^2 dx$

$(C)\pi \int\limits_{0}^{7.24264} \left(\frac{1}{2}\left(-x^2+6x+9\right)\right)^2 dx$

$(D)\dfrac{1}{2} \int\limits_{-1.24264}^{7.24264} \left(-x^2+6x+9\right)^2 dx$

Answer _____

43. Let R be the region that lies inside the polar curve $r = 2 - \cos\theta$ and also inside the polar curve $r = 1 + \cos\theta$. What is the area of R?

(A)1.611 $\quad$ (B)2.658 $\quad$ (C)2.457 $\quad$ (D)3.222

Answer_____

44. What is the volume generated when the region bounded by $y = \tan x, y = \cos x,$ and the y-axis is revolved about the x-axis?

(A) 0.792 $\quad$ (B)1.186 $\quad$ (C) 1.246 $\quad$ (D) 1.433

Answer _____

45. The amount y of a radioactive substance decays according to the equation $\dfrac{dy}{dt} = ky$ where k is a constant and time, t, is measured in days. If half of the amount present will decay in 13 days, what is the value of k?

(A) -0.053

(B) -0.015

(C) 0.053

(D) 0.015

Answer_____

Examination VI

Section I Part A

Directions: Solve each of the following problems, using the space provided. Choose the best answer. Do not spend too much time on any one problem. Calculators may NOT be used on this part of the exam.

In this test: Unless otherwise specified, the domain of a function is assumed to be the set of all real numbers x for which $f(x)$ is a real number.

1. $\lim\limits_{x \to 0} \dfrac{\sin x + 2x}{4x}$ is

$(A)\ \dfrac{1}{2}$

$(B)\ \dfrac{3}{4}$

$(C)\ \dfrac{3}{8}$

(D) Nonexistent

Answer_____

2. Which of the following is the second derivative of the function described by the pair of parametric equations $x = \sin t$ and $y = \cos t$?

$(A) \sec^3 t$ $(B) -\sec^3 t$ $(C) -\dfrac{\sec^2 t}{\sin t}$ $(D) -\sec^2 t$

Answer _____

3. If $f(x) = e^e$, which of the following is $f'(x)$?

(A) e^e

(B) 0

(C) e

(D) e^{e-1}

Answer_____

4. If $f(x) = x^3 - 6x$, which of the following are the absolute maximum and absolute minimum of $f(x)$ on the interval $[-3,2]$?

(A) Max: $f(\sqrt{2})$, Min: $f(-\sqrt{2})$　　(B) Max: $f(\sqrt{2})$, Min: $f(-3)$

(C) Max: $f(-\sqrt{2})$, Min: $f(2)$　　(D) Max: $f(-\sqrt{2})$, Min: $f(-3)$

Answer _____

5. Given the function $f(x) = x^2 + 4x - 1$, for which of the following values of c on the open interval $(0,5)$ will the conclusion of the Mean Value theorem be satisfied for the function $f(x)$?

(A) $\dfrac{9}{2}$

(B) 4

(C) $\dfrac{5}{2}$

(D) $\dfrac{1}{3}$

Answer_____

6. $\left(\tan^{-1}\left(e^{x^2} \right) \right)' = ?$

(A) $\dfrac{2xe^{x^2}}{1+e^{2x^2}}$ (B) $\dfrac{4xe^{x^2}}{1+e^{x^4}}$ (C) $\dfrac{2x}{1+e^{2x^2}}$ (D) $\dfrac{2xe^{x^2}}{1+e^{x^2}}$

Answer_____

7. $\int \dfrac{x^2-2}{x^3-6x+1}dx =$

(A) $\ln|x^3-6x+1|+C$

(B) $\dfrac{1}{3}\ln|x^3-6x+1|+C$

(C) $\dfrac{1}{3}(x^3-6x+1)+C$

(D) $\dfrac{(x^3-6x+1)^2}{2}+C$

Answer _____

8. The region is bounded by the curve of $y=\sin^2 x$ (graph of which is shown below) and the $x-$axis from $x=0$ to $x=\pi$. If this region is revolved around the x-axis, which of the following is the volume of the resulting solid?

(A) $\dfrac{3}{8}\pi$ (B) $\dfrac{1}{2}\pi^2$ (C) $\dfrac{3}{8}\pi^2$ (D) $\dfrac{3}{4}\pi^2$

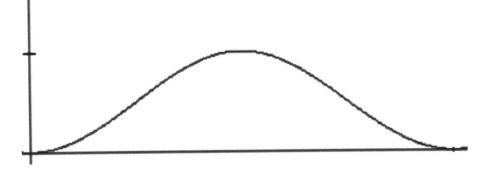

Answer _____

9. Let $\dfrac{dy}{dx} = e^{x-y}$. Which of the following is the solution to this equation such that $y(0) = 1$?

$(A)\ y = \ln x$

$(B)\ y = \ln\left(e^x + e\right)$

$(C)\ y = e^x$

$(D)\ y = \ln\left(e^x + e - 1\right)$

Answer_____

10. A wire 80 inches long is to be cut into two pieces. One piece is to be bent into the shape of a circle and the other into a square. What should be the length of the smaller piece so that the sum of the areas of the circle and the square is minimized?

$(A)\ \dfrac{80\pi}{8 + 2\pi}$ $(B)\ \dfrac{80\pi}{4 + \pi}$ $(C)\ \dfrac{160}{4 + \pi}$ $(D)\ \dfrac{80}{4 + \pi}$

Answer _____

11. $\int \dfrac{18x-17}{(2x-3)(x+1)}dx =$

$(A)\, 8\ln|2x-3|+7\ln|x+1|+C$

$(B)\, 2\ln|2x-3|+7\ln|x+1|+C$

$(C)\, 4\ln|2x-3|+7\ln|x+1|+C$

$(D)\, 7\ln|2x-3|+2\ln|x+1|+C$

Answer_____

12. What is the area of the region R enclosed by the graphs of $y = \ln(x^2), x = 1$ and $x = 3$?

$(A)\, 6\ln 3 - 4$ $(B)\, 3\ln 3 - 2$ $(C)\, 3\ln 3 + 2$ $(D)\, 3\ln 3 - 3$

Answer_____

13. Which of the following is the power series representation for the function $f(x) = \dfrac{3}{(1+x)^2}$ if $|x| < 1$?

(A) $3 - 3x + 3x^2 - 3x^3 + \ldots$

(B) $1 - x + x^2 - x^3 + \ldots$

(C) $3 - 3(2x) + 3(3x^2) - 3(4x^3) + \ldots$

(D) $1 - 2x + 3x^2 - 4x^3 + \ldots$

Answer _____

14. A curve is given parametrically by $x = \ln t, y = e^t$. What is the equation of the line tangent to the curve at $t = 12$?

(A) $y - e = 12e^{12}\left(x - e^{12}\right)$

(B) $y - e^{12} = 12e^{12}\left(x - \ln 12\right)$

(C) $y + e^{12} = -12e^{12}\left(x + \ln 12\right)$

(D) $y - e^{12} = e^{12}\left(x - \ln 12\right)$

Answer _____

15. Which of the following series converge?

I. $\sum\limits_{n=2}^{\infty} \dfrac{1}{n^2 \cdot \ln n}$　　II. $\sum\limits_{n=2}^{\infty} \dfrac{1}{n \cdot \ln n}$　　III. $\sum\limits_{n=2}^{\infty} \dfrac{1}{\sqrt{n} \cdot \ln n}$

(A) I only

(B) II only

(C) I and II only

(D) I, II, and III

Answer _____

16. The position $s(t)$ of a particle on the x-axis at time t, $t \geq 0$, is $\cos t$. The average velocity of the particle for $0 \leq t \leq \pi$ is

$(A) -\dfrac{1}{\pi}$　$(B) \dfrac{2}{\pi}$　$(C) -\dfrac{2}{\pi}$　$(D) \dfrac{1}{\pi}$

Answer _____

x	0	4	6	8	16
$g(x)$	28	27	24	19	12
$g'(x)$	0	-1	-2	-3	-4

17. Let $g(x)$ be a differentiable function defined on the interval $0 \le x \le 16$. Some values of $g(x)$ and its derivative $g'(x)$ are given in the table above. Which of the following is the x - intercept of the line tangent to the graph of $g(x)$ and parallel to the segment connecting the endpoints of $g(x)$?

$(A)\ (23,0)$

$(B)\ (31,0)$

$(C)\ (-23,0)$

$(D)\ (-31,0)$

Answer_____

18. $\lim\limits_{x \to 0} \dfrac{\displaystyle\int_{2}^{2+x} \dfrac{\sin t}{3t+2}}{x} =$

$(A)\dfrac{\cos 2}{2}$ $(B)\dfrac{\sin 2}{4}$ $(C)\dfrac{\sin 2}{8}$ $(D)\dfrac{\sin(2+x)}{x(3x+2)}$

Answer_____

Examination VI

19. Which of the following is the power series representation for $f(x) = x \cdot (\text{Arc} \tan x)$ if $|x| < 1$?

(A) $1 - x + \dfrac{x^3}{3} - \dfrac{x^5}{5} + \dfrac{x^7}{7} - \dfrac{x^9}{9} + \dots$

(B) $1 + x + \dfrac{x^3}{3} + \dfrac{x^5}{5} + \dfrac{x^7}{7} + \dfrac{x^9}{9} + \dfrac{x^{11}}{11} + \dots$

(C) $x^2 - \dfrac{x^4}{3} + \dfrac{x^6}{5} - \dfrac{x^8}{7} + \dfrac{x^{10}}{9} + \dots$

(D) $x - \dfrac{x^3}{3} + \dfrac{x^5}{5} - \dfrac{x^7}{7} + \dfrac{x^9}{9} + \dots$

Answer_____

20. $\dfrac{d}{dx}\displaystyle\int_0^{x^3} \sqrt{t^2 + 2}\, dt =$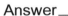

(A) $\sqrt{x^6 \cdot 3x^2 + 2} - \sqrt{2}$

(B) $\sqrt{x^6 + 2}$

(C) $\left(\sqrt{x^6 + 2}\right) \cdot 3x^2$

(D) $\left(\sqrt{x^6 + 2}\right) \cdot x^3$

Answer_____

Examination VI

21. If the length of a curve $y = f(x)$ from $x = a$ to $x = b$ is given by $L = \int_a^b \sqrt{x^2 + 2x + 2}\ dx$ then $f(x)$ could be

(A) $\dfrac{x^2}{2} + 2x$

(B) $(x+1)^2$

(C) $\dfrac{x^2}{2} + x + 3$

(D) $\dfrac{x^2}{2} + 3x + 2$

Answer_____

22. Given f is continuous for all real numbers, $\dfrac{dy}{dx} = f(x)$ and $y(-1) = 6$, what is $y(x)$?

(A) $6 + \int_{-1}^{x} f(t)\,dt$

(B) $\int_{-1}^{x} f(t)\,dt - 6$

(C) $6 + \int_{-1}^{x} f'(t)\,dt$

(D) $6 - \int_{-1}^{x} f'(t)\,dt$

Answer_____

23. $\int \dfrac{\cos 2x}{\sin^2 x \cdot \cos^2 x} dx =$

$(A) - \tan x + \cot x + C$

$(B) - \tan x - \cot x + C$

$(C) \tan x - \cot x + C$

$(D) \ln|\csc x - \cot x| - \ln|\sec x + \tan x| + C$

Answer_____

24. A spherical balloon is being filled with air. The radius increases at a rate of 1 inch per second. What is the instantaneous rate of change in volume when the radius of the balloon is 8 inches?

$(A)\ 36\pi$

$(B)\ 64\pi$

$(C)\ 72\pi$

$(D)\ 256\pi$

Answer _____

25. The region in the first quadrant bounded by the graphs of the equations $y = x^2$ and $y = 2x^2 - 4$ is revolved around the $y-$axis. What is the volume of the solid of revolution?

(A) 2π $\qquad$ (B) 3π $\qquad$ (C) 4π $\qquad$ (D) 5π

Answer _____

26. Which of the following is an equation of the normal line to the graph of $y = \dfrac{1 - \cos x}{1 + \cos x}$ at the

point $P\left(\dfrac{\pi}{2}, 1\right)$?

(A) $y - 1 = -\dfrac{1}{2}\left(x - \dfrac{\pi}{2}\right)$

(B) $y - 1 = 2\left(x - \dfrac{\pi}{2}\right)$

(C) $y - 1 = \dfrac{1}{2}\left(x - \dfrac{\pi}{2}\right)$

(D) $y - 1 = -2\left(x - \dfrac{\pi}{2}\right)$

Answer _____

Examination VI

27. $\lim\limits_{h \to 0} \dfrac{4\cos^4(x+h)+3\sin(x+h)-4\cos^4 x-3\sin x}{h}$ is

$(A)\ 0$

$(B)\ 16\cos^3 x \sin x + 3\cos x$

$(C)\ -16\cos^3 x + 3\cos x$

$(D)\ -16\cos^3 x \sin x + 3\cos x$

Answer_____

28. During the time period from $t=0$ to $t=8$, a particle moves along the path given by

$x(t)=2\sin\left(\dfrac{\pi}{2}t\right)$ and $y(t)=4\cos\left(\dfrac{\pi}{2}t\right)$. How many times does the particle pass through the

point at which $t=3$?

$(A)\ 1$ time $\quad(B)\ 2$ times $\quad(C)\ 3$ times $\quad(D)\ 4$ times

Answer_____

29. A particle moving along the polar curve given by $r = 2 + 2\sin\theta$ has position $\left(x(t), y(t)\right)$ at time t, with $\theta = 0$ when $t = 0$. This particle moves along the curve so that $\dfrac{dr}{dt} = \dfrac{dr}{d\theta}$. What is the value of $\dfrac{dr}{dt}$ at $\theta = \dfrac{\pi}{6}$?

$(A)\sqrt{3}$ $(B)\dfrac{1}{2}$ $(C)\dfrac{\sqrt{3}}{2}$ $(D)\ 1$

Answer_____

30. Two ants are moving along the curves in the xy-plane. For time $t \geq 0$, the position of ant A is given by $x = 2t - 4$ and $y = (2t - 4)^2$, and the position of ant B is given by $x = \dfrac{4t}{3} - 2$ and $y = \dfrac{4t}{3}$. What is the the exact time t at which the two ants meet each other?

$(A)\ t = 2$ $(B)\ t = \dfrac{4}{3}$ $(C)\ t = 3$ $(D)t = \dfrac{5}{3}$

Answer_____

Examination VI

Directions : Solve each of the following problems, using the space provided. Choose the best answer. Do not spend too much time on any one problem. A graphing calculator is required for some questions on this part of the exam.

In this test :

1. The exact numerical value of the correct answer does not always appear among the choices given. Then select from among the choices the number that best approximates the exact numerical value.

2. Unless otherwise specified, the domain of a function is assumed to be the set of all real numbers x for which $f(x)$ is a real number.

31. $f(0.6)$ is approximated by the fifth degree Taylor polynomial $(x-1) - \dfrac{(x-1)^3}{3!} - \dfrac{(x-1)^5}{5!}$.

Given that $\left| f^{(7)}(x) \right| \le 2$ for all x in the interval $[0.6, 1]$, what is the maximum error of $f(0.6)$?

(A) 6.89×10^{-7} (B) 6.89×10^{-8} (C) 6.50×10^{-7} (D) 6.50×10^{-8}

Answer_____

32. Water leaks from a pipe into a bucket at a rate of $1.7 + \dfrac{4.5}{x}$, x being the time in minutes that the pipe has been leaking. If there is 16.52 mL of water in the bucket after 2 minutes, how much water is in the bucket after 5 minutes?

(A) 25.74 mL (B) 24.67 mL (C) 20.12 mL (D) 16.70 mL

Answer _____

33. A cannonball is fired from the point considered to be the origin. The path of the cannonball follows the parametric equations: $y = 16\sqrt{2}t - 16t^2$ and $x = 16\sqrt{2}t$, where y is the height of the cannonball and x is the x coordinate of the cannonball. What is the total distance traveled by the cannonball from its starting point to the point at which it hits the ground again?

(A) 36.729

(B) 23.910

(C) 13.337

(D) 44.602

Answer_____

34. The function $f(x) = \left| \sin^2 x - \cos x \sin\left(\dfrac{\pi x}{2}\right) \right|$ is given on the interval $(0,5)$. Approximately at which x-coordinates is f not differentiable?

$(A)\{0.9,\ 2.4,\ 3.7\}$ $(B)\ \{0.4,\ 1.7,\ 3.1,\ 4.5\}$ $(C)\ \{2.4,\ 4.5\}$ $(D)\ \{0.9,\ 1.7,\ 4.5\}$

Answer _____

35. Let R be the region in the first quadrant bounded by the graph of $y = -4x^2 + 2kx$ and the x-axis. If the area of the region R is 18, then the value of k is

$(A)3$ $(B)6$ $(C)9$ $(D)18$

Answer_____

36. The graph of the function $f(x) = 4\cos x$ crosses the y-axis at point P $(0,4)$ and the x-axis at the point Q $\left(\dfrac{\pi}{2},0\right)$. Which of the following is the x-coordinate of the point on the graph of $f(x)$, between P and Q at which the line tangent to the graph of $f(x)$ is parallel to the segment connecting points P and Q?

(A) 0.6901

(B) 0.8901

(C) 1.0601

(D) 0.1348

Answer _____

37. What is the area in the first quadrant of the region R, bounded by the y-axis, horizontal line $y = 2$ and $y = \sqrt{x^3 + 1}$?

(A) 0.7587 (B) 1.0262 (C) 1.4423 (D) 3.2450

Answer_____

38. A ball is thrown up into the air at the velocity of 50 ft/sec. After t seconds, the distance of the ball above the ground is $50t - 16t^2$ ft. What is the velocity of the ball the instant it strikes the ground?

(A) 25 ft/sec

(B) -25 ft/sec

(C) 50 ft/sec

(D) -50 ft/sec

Answer _____

39. The number of nonzero terms of the Taylor series for $f(x) = \cos x$ about $x = 0$ that should be summed up to approximate $\cos\dfrac{3\pi}{180}$ to within 0.00001 is:

(A) 2

(B) 3

(C) 4

(C) 5

Answer _____

40. Which of the following tests prove that the series $\displaystyle\sum_{n=1}^{\infty} \dfrac{n^2}{\sqrt{n^3 - 2n}}$ diverges?

I.Nth term test   II.Limit Comparison Test with $\displaystyle\sum_{n=1}^{\infty} \dfrac{1}{n^{\frac{3}{2}}}$ III.Limit Comparison Test with $\displaystyle\sum_{n=1}^{\infty} \dfrac{1}{n}$

(A) I only (B) III only (C) I and III only (D) I, II, and III

Answer _____

41. The population $P(t)$ of a species satisfies the logistic differential equation $\dfrac{dP}{dt} = P\left(224 - \dfrac{P^2}{56}\right)$ where the initial population $P(0) = 30$ and t is the time in years. What is $\lim\limits_{t \to \infty} P(t)$?

(A) 106

(B) 112

(C) 224

(D) 12544

Answer _____

42. If the function g is defined by $g(x) = \displaystyle\int_0^{x^2} \cos t\, dt$ on the closed interval $-\dfrac{1}{2} \le x \le \dfrac{5}{2}$, then g has a local maximum at $x =$

(A) 0

(B) 0.808

(C) 1.253

(D) 2.171

Answer _____

43. At any time t $\geq$ 0, in days, the rate of a mosquito population is given by $y' = ky$, where k is a constant of proportionality and y is the number of mosquitoes present. The initial population is 3,000 and the population quadruples during the first 7 days. By what factor will the population have increased in the first 14 days?

(A) 4

(B) 8

(C) 16

(D) 32

Answer_____

44. Let R be the region enclosed by $y = 3\ln x$, $x = 5$ and $x-$ axis. What is the volume of the solid generated by revolving the region R about the x-axis?

(A) 193.879

(B) 137.330

(C) 149.532

(D) 187.913

Answer _____

45. A particle moves along the x-axis such that its acceleration from time $t=0$ to time $t=5$ is given by $a(t)=-2\sin(t)$. The particle has a velocity of 1 at time $t=0$. What is the total distance traveled by the particle over the time interval?

(A) 6.918 (B) 8.288 (C) 7.640 (D) 3.288

Answer_____

Formulas and Theorems

Limits

$\lim_{x \to c}\left[f(x)+g(x)\right] = \lim_{x \to c} f(x) + \lim_{x \to c} g(x)$	$\lim_{x \to c}\left[af(x)\right] = a \lim_{x \to c} f(x)$
$\lim_{x \to c}\left[f(x)\cdot g(x)\right] = \lim_{x \to c} f(x) \cdot \lim_{x \to c} g(x)$	$\lim_{x \to c}\left[f(x)^{\frac{a}{b}}\right] = \left[\lim_{x \to c} f(x)\right]^{\frac{a}{b}}$
$\lim_{x \to c}\left[\dfrac{f(x)}{g(x)}\right] = \dfrac{\lim_{x \to c} f(x)}{\lim_{x \to c} g(x)}$	$\lim_{n \to 0}(1+n)^{\frac{1}{n}} = \lim_{n \to \infty}\left(1+\dfrac{1}{n}\right)^{n} = e$
$\lim_{x \to 0}\dfrac{\sin x}{x} = \lim_{x \to 0}\dfrac{x}{\sin x} = 1$	$\lim_{x \to 0}\dfrac{1-\cos x}{x} = 0$

Differentiation formulas

Power Rule: $\left(x^n\right)' = nx^{n-1}$	Product Rule: $\left(f(x)\cdot g(x)\right)' = f'(x)\cdot g(x) + g'(x)\cdot f(x)$
Reciprocal Rule: $\left(\dfrac{1}{g(x)}\right)' = -\dfrac{g'(x)}{g^2(x)}$	Quotient Rule: $\left(\dfrac{f(x)}{g(x)}\right)' = \dfrac{f'(x)\cdot g(x) - g'(x)\cdot f(x)}{g^2(x)}$
$\left(\sin x\right)' = \cos x$	Chain Rule: $\left(f(g(x))\right)' = f'(g(x))\cdot g'(x)$
$\left(\cos x\right)' = -\sin x$	$\left(a\right)' = 0$
$\left(\tan x\right)' = \sec^2 x$	$\left(x\right)' = 1$
$\left(\cot x\right)' = -\csc^2 x$	$\left(e^x\right)' = e^x$
$\left(\sec x\right)' = \sec x \tan x$	$\left(\ln x\right)' = \dfrac{1}{x}$
$\left(\csc x\right)' = -\csc x \cot x$	$\left(a^x\right)' = a^x \ln a,\ a > 0,\ a \neq 1$
$\left(\sin^{-1} x\right)' = \dfrac{1}{\sqrt{1-x^2}}$	$\left(\log_a x\right)' = \dfrac{1}{x \ln a}$
$\left(\tan^{-1} x\right)' = \dfrac{1}{1+x^2}$	$\left(\sqrt{x}\right)' = \dfrac{1}{2\sqrt{x}}$
$\left(\sec^{-1} x\right)' = \dfrac{1}{x\sqrt{x^2-1}}$	$\left(\dfrac{1}{x}\right)' = -\dfrac{1}{x^2}$

Formulas and Theorems

Integration Formulas

$\int \sin x\, dx = -\cos x + C$	$\int x^n dx = \dfrac{x^{n+1}}{n+1} + C,\ n \neq -1$
$\int \cos x\, dx = \sin x + C$	$\int a\, dx = ax + C$
$\int \tan x\, dx = -\ln|\cos x| + C \text{ or } \ln|\sec x| + C$	$\int e^x\, dx = e^x + C$
$\int \cot x\, dx = \ln|\sin x| + C \text{ or } -\ln|\csc x| + C$	$\int a^x dx = \dfrac{a^x}{\ln a} + C$
$\int \sec x\, dx = \ln|\sec x + \tan x| + C$	$\int \dfrac{1}{x}\, dx = \ln|x| + C$
$\int \csc x\, dx = \ln|\csc x - \cot x| + C$	$\int \dfrac{1}{\sqrt{a^2 - x^2}}\, dx = \sin^{-1}\dfrac{x}{a} + C$
$\int \sec x \tan x\, dx = \sec x + C$	$\int \dfrac{1}{a^2 + x^2}\, dx = \dfrac{1}{a}\tan^{-1}\dfrac{x}{a} + C$
$\int \csc x \cot x\, dx = -\csc x + C$	$\int \dfrac{1}{x\sqrt{x^2 - a^2}}\, dx = \dfrac{1}{a}\sec^{-1}\dfrac{x}{a} + C$
$\int \sec^2 x\, dx = \tan x + C$	$\int \csc^2 x\, dx = -\cot x + C$

Continuity	A function $f(x)$ is continuous at $x = a$ if all of the following are true: I. $f(a)$ exists II. $\lim\limits_{x \to a} f(x)$ exists III. $\lim\limits_{x \to a} f(x) = f(a)$
Intermediate Value Theorem	If function $f(x)$ is continuous on a closed interval $[a,b]$ and if w is any number between $f(a)$ and $f(b)$, then there is at least one number c in $[a,b]$ such that $f(c) = w$
Limit Theorem	$\lim\limits_{x \to a} f(x) = L$ if and only if $\lim\limits_{x \to a^+} f(x) = L = \lim\limits_{x \to a^-} f(x)$

Formulas and Theorems

Vertical Asymptote	A line $x = a$ is a vertical asymptote of the graph of a function $y = f(x)$ if $\lim\limits_{x \to a^+} f(x) = \pm\infty$ or $\lim\limits_{x \to a^-} f(x) = \pm\infty$
Horizontal Asymptote	A line $y = L$ is a horizontal asymptote of the graph of a function $f(x)$ if $\lim\limits_{x \to \infty} f(x) = L$ or $\lim\limits_{x \to -\infty} f(x) = L$
Average Rate of Change	Average rate of change of $y = f(x)$ on $[a, a+h]$ is $\dfrac{f(a+h) - f(a)}{h}$
Instantaneous Rate of Change	Instantaneous rate of change of $y = f(x)$ on $[a, a+h]$ is: $\lim\limits_{h \to 0}(\text{Average rate of change}) = \lim\limits_{h \to 0} \dfrac{f(a+h) - f(a)}{h}$
Sandwich Theorem	If $f(x) \le g(x) \le h(x)$ for all $x \ne c$ in some interval about c, and $\lim\limits_{x \to c} f(x) = \lim\limits_{x \to c} h(x) = L$, then $\lim\limits_{x \to c} g(x) = L$
Definition of Derivative	$f'(x) = \lim\limits_{h \to 0} \dfrac{f(x+h) - f(x)}{h}$ or $f'(x) = \lim\limits_{x \to c} \dfrac{f(x) - f(c)}{x - c}$
Theorem	If a function $f(x)$ is differentiable at $x = a$, then it is continuous at $x = a$
Differentiability on a closed interval	A function f is differentiable on a closed interval $[a, b]$ if f is differentiable on the open interval (a, b) and if both Right-Hand Derivative at a and Left-Hand Derivative at b exist
Rolle's Theorem	Suppose that $f(x)$ is continuous on the closed interval $[a, b]$ and differentiable on the open interval (a, b). If $f(a) = f(b)$, then there is at least one number c between a and b such that $f'(c) = 0$
Mean Value Theorem	If $f(x)$ is continuous on the closed interval $[a, b]$ and differentiable on the open interval (a, b), then there is at least one number c between a and b such that $f'(c) = \dfrac{f(b) - f(a)}{b - a}$
Derivative of the Inverse function	$g'(c) = \dfrac{1}{f'(g(c))}$
Linear approximation of f(x) near x = x$_0$	$y = f(x_0) + f'(x_0)(x - x_0)$

Formulas and Theorems

Extrema and Concavity

Critical Numbers (Candidates for Extrema)	A number c in the domain of a function f is a critical number of f if either $f'(c)=0$ or $f'(c)$ does not exist
Increasing/Decreasing Functions	If $f(x)$ is differentiable on (a,b) and continuous on $[a,b]$: $f'(x)>0$ on $(a,b)\Rightarrow f(x)$ is increasing on $[a,b]$ $f'(x)<0$ on $(a,b)\Rightarrow f(x)$ is deccreasing on $[a,b]$
Local Minimum	$f(c)$ is the local minimum value of f if $f(c)\le f(x)$ for every x in an interval I around c
Local Maximum	$f(c)$ is the local maximum value of f if $f(c)\ge f(x)$ for every x in an interval I around c
Absolute Minimum	$f(c)$ is the absolute minimum value of f if $f(c)\le f(x)$ for every x in the domain of f
Absolute Maximum	$f(c)$ is the absolute maximum value of f if $f(c)\ge f(x)$ for every x in the domain of f
Finding the Absolute Extrema on [a,b]	1. Find Critical numbers. 2. Calculate the function values at the endpoints of $[a,b]$ and at the critical numbers. 3. The largest of these function values is the absolute maximum The smallest function value is the absolute minimum.
Finding the Local Extrema (First Derivative Test)	Find critical numbers $x=c$ $\left(\text{where } f'(c)=0 \text{ or DNE}\right)$ Local minimum occurs at $x=c$, where $f'(x)$ changes from negative to positive. Local maximum occurs at $x=c$, where $f'(x)$ changes from positive to negative .
Concavity	If $f''(x)$ exists on (a,b), then: $f''(x)>0\Rightarrow f(x)$ concave upward in (a,b) $f''(x)<0\Rightarrow f(x)$ concave downward in (a,b)
Points of Inflection	Points of Inflection of $f(x)$ are the points on the Domain of $f(x)$ where $f''(x)=0$ or DNE and $f''(x)$ changes it's sign passing through them.
Extreme Value Theorem	If f is continuous on a closed interval $[a,b]$, then f has both an absolute minimum and an absolute maximum value on $[a,b]$

Formulas and Theorems

Integrals

Definite Integral. Property 1	$\int_a^b c \cdot f(x)\,dx = c\int_a^b f(x)\,dx$
Definite Integral. Property 2	$\int_a^b f(x)\,dx = -\int_b^a f(x)\,dx$
Definite Integral. Property 3	$\int_a^a f(x)\,dx = 0$
Definite Integral. Property 4	If f is integrable on a closed interval and if a, b and c are any three numbers in the interval, then $\int_a^b f(x)\,dx = \int_a^c f(x)\,dx + \int_c^b f(x)\,dx$
If $f(x) \geq 0$ on $[a,b]$:	If $f(x) \geq 0$ on $[a,b]$ then $\int_a^b f(x)\,dx \geq 0$
If $g(x) \geq f(x)$ on $[a, b]$:	If $g(x) \geq f(x)$ on $[a, b]$ then $\int_a^b f(x)\,dx \geq 0$
If f(x) is an even function:	If f is an Even Function, then $\int_{-a}^a f(x)\,dx = 2\int_0^a f(x)\,dx$
If f(x) is an odd function:	If f is an Odd Function, then $\int_{-a}^a f(x)\,dx = 0$
Riemann sum	Let $f(x)$ be defined on a closed interval $[a,b]$ and P is any decomposition of $[a,b]$ into subintervals of the form $[x_{k-1}, x_k]$. Riemann sum of $f(x)$ for P is $R_P = \sum_{k=1}^n f(w_k)\Delta x_k$ where $w_k \in [x_{k-1}, x_k]$.
Integral as limit of Riemann sums	The definite integral of f from a to b, $\int_a^b f(x)\,dx = \lim_{\|p\| \to 0} \sum_k f(w_k)\Delta x_k$ provided the limit exists where $\|p\|$ is the norm of the partition (largest Δx_k)
Integral as area under the curve	If f is integrable and $f(x) \geq 0$ for every x in $[a,b]$ (where $a < b$), then the area A of the region under the graph of f from a to b is $A = \int_a^b f(x)\,dx$
Fundamental Theorem of Calculus	$\dfrac{d}{dx}\int_a^x f(t)\,dt = f(x)$ and $\int_a^b f'(x)\,dx = f(b) - f(a)$
Velocity	$v(t) = x'(t)$ $v(t) = \int a(t)\,dt$
Acceleration	$a(t) = v'(t) = x''(t)$

Formulas and Theorems

Position	$x(t) = \int v(t)\,dt$
Average Velocity	$v_{av} = \dfrac{s(t_2) - s(t_1)}{t_2 - t_1}$
Instantaneous velocity	$v_{inst} = \lim\limits_{t_2 \to t_1} \dfrac{s(t_2) - s(t_1)}{t_2 - t_1}$
Total Distance	Total Distance $= \int_a^b |v(t)|\,dt$
Net Change	Net Change $= \int_a^b v(t)\,dt$
Area between the curves	$A = \int_a^b \big(f(x) - g(x)\big)\,dx$, where $f(x) > g(x)$ on $[a,b]$
Volume by Disks	$V = \pi \int_a^b R^2\,dx$ (if region is rotating around horizontal line) or: $V = \pi \int_c^d R^2\,dy$ (if region is rotating around vertical line)
Volume by Washers	$V = \pi \int_a^b \big(R^2 - r^2\big)\,dx$ (if region is rotating around horizontal line) or: $V = \pi \int_c^d \big(R^2 - r^2\big)\,dy$ (if region is rotating around vertical line)
Volume by Cross Sections	$V = \int_a^b A(x)\,dx$ (if the cross sections are $\perp$ to x-axis) or: $V = \int_c^d A(y)\,dy$ (if the cross sections are $\perp$ to y-axis)
Average value of f	Average Value of $f(x)$ on $[a,b]$ is: $f_{av} = \dfrac{1}{b-a}\int_a^b f(x)\,dx$
Trapezoidal Rule	$\int_a^b f(x)\,dx = \dfrac{b-a}{2n}\big[f(a_0) + 2f(a_1) + 2f(a_2) + \ldots + f(a_n)\big]$
Power-reducing trig formulas	$\sin^2 x = \dfrac{1 - \cos 2x}{2}$ and $\cos^2 x = \dfrac{1 + \cos 2x}{2}$
Arc Length	Arc Length: $L = \int_a^b \sqrt{1 + \big[f'(x)\big]^2}\,dx$ or: $L = \int_c^d \sqrt{1 + \big[g'(y)\big]^2}\,dy$

Formulas and Theorems

Integration by parts	$\int u\,dv = uv - \int v\,du$
L'Hopital's Rule	If $\lim\limits_{x\to c}\dfrac{f(x)}{g(x)}$ is of the form $\dfrac{0}{0}$ or $\dfrac{\infty}{\infty}$, then $\lim\limits_{x\to c}\dfrac{f(x)}{g(x)} = \lim\limits_{x\to c}\dfrac{f'(x)}{g'(x)}$, provided either $\lim\limits_{x\to c}\dfrac{f'(x)}{g'(x)}$ exists or $\lim\limits_{x\to c}\dfrac{f'(x)}{g'(x)} = \infty$
Improper Integrals	If $f(x)$ is continuous on $[a,\infty)$, then $\int_a^\infty f(x)\,dx = \lim\limits_{t\to\infty}\int_a^t f(x)\,dx$ If $f(x)$ is continuous on $[-\infty,a)$, then $\int_{-\infty}^a f(x)\,dx = \lim\limits_{t\to-\infty}\int_t^a f(x)\,dx$
Improper Integrals	If $f(x)$ is continuous on $(a,b]$ and discontinuous at a, then $\int_a^b f(x)\,dx = \lim\limits_{t\to a^+}\int_t^b f(x)\,dx$ If $f(x)$ is continuous on $[a,b)$ and discontinuous at b, then $\int_a^b f(x)\,dx = \lim\limits_{t\to b^-}\int_a^t f(x)\,dx$ If $f(x)$ has discontinuity at a number c in the interval (a,b), but continuous elsewhere on $[a,b]$, then $\int_a^b f(x)\,dx = \int_a^c f(x)\,dx + \int_c^b f(x)\,dx$

Parametric Equations and Polar coordinates

Derivative of a function given parametrically	$\dfrac{dy}{dx} = \dfrac{\frac{dy}{dt}}{\frac{dx}{dt}}$
Second derivative of a function given parametrically	$\dfrac{d^2y}{dx^2} = \dfrac{\frac{dy'}{dt}}{\frac{dx}{dt}}$
Arc Length of a curve given parametrically	$L = \int_a^b \sqrt{\left[f'(t)\right]^2 + \left[g'(t)\right]^2}\,dt = \int_a^b \sqrt{\left(\dfrac{dx}{dt}\right)^2 + \left(\dfrac{dy}{dt}\right)^2}\,dt$
Polar to Cartesian	$x = r\cos\theta \quad y = r\sin\theta$
Cartesian to Polar	$x^2 + y^2 = r^2 \quad \tan\theta = \dfrac{y}{x}$
Area in polar coordinates	If f is continuous and $f(\theta) \geq 0$ on $[\alpha,\beta]$, where $0 \leq \alpha < \beta \leq 2\pi$, then the area A of the region bounded by the graphs of $r = f(\theta)$, $\theta = \alpha$, and $\theta = \beta$ is $$A = \frac{1}{2}\int_\alpha^\beta r^2\,d\theta = \frac{1}{2}\int_\alpha^\beta (f(\theta))^2\,d\theta$$
Area between the curves	$A = \dfrac{1}{2}\int_\alpha^\beta \left(r_2^2 - r_1^2\right)d\theta$

Formulas and Theorems

Vectors

Position: $s(t) = \langle x(t), y(t) \rangle$

Velocity: $v(t) = s'(t) = \left\langle \dfrac{dx}{dt}, \dfrac{dy}{dt} \right\rangle$

Speed: $|v(t)| = \sqrt{\left(\dfrac{dx}{dt}\right)^2 + \left(\dfrac{dy}{dt}\right)^2}$

Acceleration: $a(t) = v'(t) = s''(t) = \left\langle \dfrac{d^2x}{dt^2}, \dfrac{d^2y}{dt^2} \right\rangle$

Sequences and Series

Sequence	If a sequence $\{a_n\}$ has a limit L $\left(\lim\limits_{n \to \infty} a_n = L\right)$, then a sequence is said to converge to L. If there is no limit, the sequence diverges.
Sandwich Theorem	If $\{a_n\}, \{b_n\}$ and $\{c_n\}$ are sequences and $a_n \le b_n \le c_n$ for every n and if $\lim\limits_{n \to \infty} a_n = L = \lim\limits_{n \to \infty} c_n$, then $\lim\limits_{n \to \infty} b_n = L$
Theorem	Series $\sum a_n$ converges if sequence of partial sums $\{S_n\}$ converges
Harmonic Series	The Harmonic Series $\sum\limits_{n=1}^{\infty} \dfrac{1}{n} = 1 + \dfrac{1}{2} + \dfrac{1}{3} + \ldots + \dfrac{1}{n} + \ldots$ diverges
Geometric Series	Geometric Series : $a + ar + ar^2 + \ldots + ar^n + \ldots \quad (a \ne 0)$ I. Converges and has the sum $S = \dfrac{a}{1-r}$ if $|r| < 1$ II. Diverges if $|r| \ge 1$
Nth Term Test for divergence	(I) If $\lim\limits_{n \to \infty} a_n \ne 0$, the series $\sum a_n$ is divergent. (II) If $\lim\limits_{n \to \infty} a_n = 0$, investigate further
Theorem	For any positive integer k, the series $\sum\limits_{n=1}^{\infty} a_n = a_1 + a_2 + \ldots$ and $\sum\limits_{n=k+1}^{\infty} a_n = a_{k+1} + a_{k+2} + \ldots$ either both converge or both diverge

Formulas and Theorems

Theorem	If $\sum a_n$ and $\sum b_n$ are convergent series with sums A and B, respectively, then I. $\sum (a_n + b_n)$ converges and has a sum $A + B$ II. $\sum ca_n$ converges and has a sum cA for every real number c III. $\sum (a_n - b_n)$ converges and has a sum $A - B$
Theorem	If $\sum a_n$ is a convergent series and $\sum b_n$ is divergent, then $\sum (a_n + b_n)$ is divergent
Integral Test	Suppose $f(x)$ is a continuous, positive, decreasing function on $[1, \infty)$ and let $a_n = f(n)$. Then the series $\sum\limits_{n=1}^{\infty} a_n$ and the improper integral $\int\limits_{1}^{\infty} f(x)\,dx$ both converge or both diverge.
P-Series	The p-series $\sum\limits_{n=1}^{\infty} \dfrac{1}{n^p} = 1 + \dfrac{1}{2^p} + \dfrac{1}{3^p} + ... + \dfrac{1}{n^p} + ...$ converges if $p > 1$ and diverges if $p \leq 1$
Basic Comparison Test	Let $\sum a_n$ and $\sum b_n$ be positive-term series. If $\sum b_n$ converges and $a_n \leq b_n$ then $\sum a_n$ also converges. If $\sum b_n$ diverges and $a_n \geq b_n$ then $\sum a_n$ also diverges.
Limit Comparison Test	Let $\sum a_n$ and $\sum b_n$ be positive-term series. If $\lim\limits_{n \to \infty} \dfrac{a_n}{b_n} = c > 0$ then either both series converge or both diverge.
Alternating Series Test	The alternating series $\sum\limits_{n=1}^{\infty} (-1)^{n-1} a_n = a_1 - a_2 + a_3 - ... + (-1)^{n-1} a_n + ...$ is convergent if I. $a_k \geq a_{k+1} > 0$ for every k II. $\lim\limits_{n \to \infty} a_n = 0$
Error Theorem for Alternating series	Let $\sum\limits_{n=1}^{\infty} (-1)^{n-1} a_n$ be Convergent Alternating Series. If S is the sum of the series and S_n is a partial sum, then $$|S - S_n| \leq a_{n+1}$$ that is, the error involved in approximating S by S_n, $|E| \leq a_{n+1}$
Absolutely Convergent Series	$\sum a_n$ is absolutely convergent if the series $\sum |a_n| = |a_1| + |a_2| + ... + |a_n| + ...$ is convergent
Conditionally Convergent Series	$\sum a_n$ is conditionally convergent if $\sum a_n$ is convergent and $\sum |a_n|$ is divergent.
Theorem	If a series $\sum a_n$ is absolutely convergent, then $\sum a_n$ is convergent.

Formulas and Theorems

Ratio Test	Let $\sum a_n$ be a series with non-zero terms and suppose $\lim_{n \to \infty} \left| \dfrac{a_{n+1}}{a_n} \right| = L$ I. If $L < 1$, series converges absolutely II. If $L > 1$ or $\lim_{n \to \infty} \left| \dfrac{a_{n+1}}{a_n} \right| = \infty$, series diverges III. If $L = 1$, then the test is inconclusive and different test must be used.
Power Series in x	A power series in x is a series of the form $\sum_{n=0}^{\infty} a_n x^n = a_0 + a_1 x + a_2 x^2 + \dots + a_n x^n + \dots,$ where each a_k is a real number.
Power Series in x-c	A power series in $x - c$ is a series of the form $\sum_{n=0}^{\infty} a_n (x-c)^n = a_0 + a_1 (x-c) + a_2 (x-c)^2 + \dots + a_n (x-c)^n + \dots,$ where each a_k and c are real numbers.
Interval of convergence	The set of all numbers x for which the power series converges is called the interval of convergence
Frequently Used Series	$\dfrac{1}{1-x} = 1 + x + x^2 + x^3 + \dots + x^n + \dots = \sum_{n=0}^{\infty} x^n, \ |x| < 1$ $e^x = 1 + x + \dfrac{x^2}{2!} + \dfrac{x^3}{3!} + \dots + \dfrac{x^n}{n!} + \dots = \sum_{n=0}^{\infty} \dfrac{x^n}{n!}, \ |x| < \infty$ $\sin x = x - \dfrac{x^3}{3!} + \dfrac{x^5}{5!} - \dfrac{x^7}{7!} + \dots + (-1)^n \dfrac{x^{2n+1}}{(2n+1)!} + \dots = \sum_{n=0}^{\infty} \dfrac{(-1)^n x^{2n+1}}{(2n+1)!}, \ |x| < \infty$ $\cos x = 1 - \dfrac{x^2}{2!} + \dfrac{x^4}{4!} - \dots + \dfrac{(-1)^n x^{2n}}{(2n)!} + \dots = \sum_{n=0}^{\infty} \dfrac{(-1)^n x^{2n}}{(2n)!}, \ |x| < \infty$
Maclaurin Series or Taylor Series at x=0	$f(x) = \sum_{n=0}^{\infty} \dfrac{f^{(n)}(0)}{n!} x^n = f(0) + f'(0)x + \dfrac{f''(0)}{2!}x^2 + \dfrac{f'''(0)}{3!}x^3 + \dots + \dfrac{f^{(n)}(0)}{n!}x^n + \dots$ where f is the function which has derivatives of all orders throughout some interval containing $x = 0$.
Taylor Series at x=a	$f(x) = \sum_{n=0}^{\infty} \dfrac{f^{(n)}(a)}{n!} (x-a)^n = f(a) + f'(a)(x-a) + \dfrac{f''(a)}{2!}(x-a)^2 + \dots + \dfrac{f^{(n)}(a)}{n!}(x-a)^n + \dots$ where f is the function which has derivatives of all orders throughout some interval containing $x = a$.
Lagrange Error Bound	$\text{Error} \le |R_n(x)| = \left| \dfrac{f^{(n+1)}(c)}{(n+1)!} (x-a)^{n+1} \right|$ where c is between x and a.

Tips for the AP Test

- Functions could have various notations such as: $Si(x), erf(x)$

- Don't cross out your work unless you know you can do better. But if you make a mistake, cross it out. A crossed out solution is not graded.

- If you are afraid your result is wrong in Part A, use it anyway to finish the problem.

- Don't write $f(x) = 2(1.5) + 3$ when you mean $f(1.5) = 2(1.5) + 3$.

- Name the function you are referring to. Not "its slope is…" but "the slope of g is…" when more than one function is being used.

- If you are being asked to set up an integral and you are not sure about integrand, start with at least the limits of integration and constant in front of it and make a guess at the integrand.

- Know the difference between local (relative) and global (absolute) extrema.

- Know the difference between the **extreme value** and the **location** of the extreme value. If they ask you to "find the *minimum*", they want the *y-value*. If they ask, "At which *x* does this function have a minimum?" they want the *x-value*. If they ask, "At which *point* does this function have a minimum?" they want *both coordinates (x,y)*.

- Know the difference between a **point in time** and an **interval of time**. "During, before", "within", "after", and "until" are intervals. "At" or "when" indicate a point in time!

- If you use a rule, show how it applies to the given problem. Generic theorems or procedures get no credit. Don't write "By Mean Value Theorem" etc.; instead explain how the conditions and rules apply to your problem.

- Volume problems that are given **will not** require use of the **Shell method**. Shell method can be used, but problem could also be solved using Disc, Washers, or Cross Section method.

- If in Washer method you write $(f(x) - g(x))^2$ vs. $(f(x))^2 - (g(x))^2$ you might **LOSE ALL POINTS & NOT BE GIVEN ANY PARTIAL CREDIT.**

- In rotation problems, when the solid is rotating around a Horizontal line *y* = 2, or Vertical line *x* = 1, be careful about rotating the figure around the **correct** line.

- Answers have to be reasonable. If you make a mistake in finding the area under the curve, and the answer is negative, it is UNREASONABLE. You can lose points for such errors and NOT BE GIVEN PARTIAL CREDIT.

- If you have "=" sign make sure that left is equal to the right. Don't put something like $2 + 5 = 7 + 1 = 8$. Instead put $2 + 5 = 7; 7 + 1 = 8$.

- Point-slope form of an equation, $y - y_0 = m(x - x_0)$, is perfectly acceptable. There is no need to simplify it into $y = mx + b$ form. The form $\dfrac{y - y_0}{x - x_0} = m$ is not acceptable because of the restrictions for the denominator.

- Answers like $2 + \sin\dfrac{\pi}{3}$, $2 + \tan^{-1}1$, $\dfrac{(.1)^2 x^2}{2!} + \dfrac{(.1)^3 x^3}{3!}$ are acceptable. You don't have to simplify it anymore.

- When you use trapezoidal rule, your answer is a complicated sum. Leave it as a sum, like 2.71 + 3.06 + 5.06. You don't have to do the arithmetic and add the terms up.

- THEY DON'T ACCEPT diagrams for justification of local maximum, local minimum, and concavity. Instead you need to write the explanation similar to: **Local minimum occurs at x = 3, because that's where derivative of** f (or g or h etc.) **changes from negative to positive.**

- For every part of the free response, give all the answers with at least 3 digits after the decimal point, rounded or truncated. More is okay. Make sure the numbers you are using for earlier steps of the solution contain MORE THAN 3 digits after the decimal point. Don't round too soon, or else your answer may be off.

- When graphing the slope field (direction fields), extend it all the way through the given window.

Answer Key

Exam 1	Exam 2	Exam 3	Exam 4	Exam 5	Exam 6
1) C	1) C	1) C	1) B	1) D	1) B
2) B	2) C	2) B	2) D	2) D	2) B
3) B	3) C	3) C	3) D	3) D	3) B
4) C	4) B	4) A	4) D	4) D	4) D
5) A	5) D	5) A	5) C	5) D	5) C
6) B	6) D	6) A	6) A	6) D	6) A
7) A	7) B	7) A	7) C	7) C	7) B
8) C	8) D	8) B	8) C	8) A	8) C
9) C	9) C	9) A	9) C	9) D	9) D
10) B	10) B	10) C	10) B	10) C	10) B
11) A	11) D	11) A	11) A	11) A	11) B
12) B	12) D	12) B	12) D	12) C	12) A
13) C	13) C	13) D	13) B	13) B	13) C
14) C	14) A	14) B	14) A	14) A	14) B
15) C	15) D	15) A	15) A	15) A	15) A
16) A	16) C	16) C	16) A	16) C	16) C
17) C	17) A	17) A	17) B	17) C	17) B
18) A	18) B	18) C	18) D	18) D	18) C
19) C	19) C	19) D	19) C	19) B	19) C
20) D	20) C	20) A	20) C	20) D	20) C
21) D	21) B	21) C	21) B	21) B	21) C
22) B	22) C	22) A	22) D	22) B	22) A
23) A	23) B	23) A	23) C	23) B	23) B
24) B	24) B	24) D	24) A	24) A	24) D
25) B	25) D	25) C	25) B	25) B	25) C
26) C	26) D	26) D	26) A	26) C	26) A
27) B	27) C	27) D	27) B	27) B	27) D
28) B	28) B	28) A	28) C	28) C	28) B
29) B	29) B	29) A	29) A	29) C	29) A
30) B	30) D	30) B	30) D	30) B	30) C
31) C	31) C	31) D	31) A	31) B	31) C
32) A	32) B	32) C	32) C	32) B	32) A
33) C	33) C	33) A	33) C	33) B	33) A
34) B	34) C	34) A	34) D	34) A	34) A
35) D	35) A	35) B	35) A	35) C	35) B
36) C	36) B	36) B	36) C	36) C	36) A
37) C	37) B	37) B	37) D	37) B	37) B
38) A	38) C	38) A	38) C	38) C	38) D
39) C	39) C	39) C	39) C	39) B	39) A
40) D	40) D	40) A	40) D	40) D	40) A
41) B	41) D	41) C	41) D	41) A	41) B
42) D	42) B	42) A	42) A	42) B	42) C
43) B	43) C	43) A	43) B	43) B	43) C
44) B	44) B	44) C	44) C	44) D	44) B
45) C	45) C	45) B	45) C	45) A	45) B

Made in the USA
Lexington, KY
20 April 2019